All about the Personal Computer

Carlos E. Hattab

First Edition

1stBooks-rev. 5/04/01

To my children: John, Wardeh and Noora

You gave me the inspiration to write this book

I love you, ...

Sincere and special thanks to:

Linda Hattab

Jo Anne Reno

Grace Melvin

Philip O'Halloran

Michael Powell

for proofreading this book.

TABLE OF CONTENTS

- *General Protection Fault Error (GPF Error)*
- *Bitmap*
- *Joint Photographic Experts Group (JPEG)*
- *Progressive JPEG*
- *Tag Image File Format (TIFF)*
- *Graphical User Interface (GUI)*
- *Graphics Interchange Format (GIF)*
- *GIF89a*
- *Animated GIF*
- *Portable Network Graphics (PNG)*
- *interlaced GIF*
- *Raster Graphics*
- *Vector Graphics*
- *Software bug*
- *Computer crash*
- *Software Program*

ABSTRACT

When shopping for or thinking about upgrading a computer, most people come across this thought "*I am not that comfortable making a purchase yet because I don't know which computer is best for me*".

I have been working with computers since 1978. I have used *CP/M, DM5, OS/2, VMS, DOS, UNIX* and many other custom *Operating Systems*. These computers were mostly in offices. They have advantages and disadvantages but one thing has changed; the Personal Computer (PC) at home. After constructing, repairing and upgrading thousands of PCs over the years, I have decided to write "***All about the Personal Computer***" to help the novice and advanced computer users understand the PC and its components. Many computer terminologies have been created in the past years, some we remember, most we forget, this book is a refresher of those "buzz words".

The *Personal Computer* is used today most everywhere: at home, to play games, check investments, access the Internet; at the office, to create a document, forecast marketing sales, issue an invoice; at school, to assist a teacher with grading, help a student with math and provide a librarian books database; at restaurants, to write in a customer order, or check supply inventory; etc.

Computers can value the quality of information. The value of information may vary depending upon **what, how, where, when** and **why** the data was gathered, processed and stored. The most valued information needs to be:

- *Timely:* Information sometimes loses its value as it ages. Different people have different needs of that information, i.e., yesterday's news is not as important as current events.

- *Accurate:* Information should be communicated without errors. The degree of accuracy depends on different circumstances. The inaccuracies can occur when data is entered into the computer system, or it can be inappropriately processed. Erroneous data put into the system brings about a situation known as "garbage in – garbage out" or "GIGO".

- *Verifiable*: The accuracy of the information entered into the computer can be confirmed, re-entered or audited. Information that cannot be verified is difficult to depend on for decision-making.

- *Relevant*: Information should enable decisions to be made easier and successfully. Extraneous information can complicate decision-making. Even the timeliest information is useless if it does not contribute to making the proper decision.

- *Complete*: Information may be meaningless if it is not complete. Circumstances may help determine if more information is needed. Accurate, verifiable, timely and relevant information is meaningless if it is incomplete.

- *Clear*: Information should not contain any ambiguous terms. It must be clear and leave no doubt concerning the meaning of facts. A generated information report that contains vague generalities and ambiguous terms quickly loses its value to the reader.

This book is intended for the novice and advanced computer users, it is about:

o *what* is it: the definition.
o *how* does it function within a computer: the summary process.
o *when* is it safe to restore: backup procedures.
o *where* does it work: its place inside the computer.
o *why* was it given a name: a little story.
o *who* discovered it: background history.

INTRODUCTION

Floppy disk, Pentium, Megabytes, Megahertz, CD ROM, E-mail … these words are related to computers. When computers became commercially available, large businesses were the only ones that could acquire them for specific and limited use (payroll, research, etc…). Until the late 1970's, computers were tools for the "specially skilled" scientists, mathematicians and engineers. That was changed in the early 1980's with the Personal Computer. That computer invaded the small businesses, the homes and the schools. By the 1990's, most people (globally) were using computers for daily activities and/or owned (at home) at least a computer.

Computers have come a long way, technology is still evolving and challenges are still being pursued by the computer industry. In 1944, the *Mark I*, an electromechanical computer, could multiply two numbers in 6 seconds and divide two numbers in about 12 seconds. In 1994, computers could perform well over 100 million additions per second. Modern computers are small in size and can run with low power (such as notebook and palm computers). In short, since the introduction of the *ENIAC* (the first and largest known computer in the world), Computer system design has evolved rapidly to about five generations:

1. Vacuum tubes – Very Large digital computers (1945 – 1958).
2. Transistors – Mainframe computers (1959 – 1964).
3. Integrated Circuits (IC) – Minicomputers (1965 – 1970).
4. LSI, VLSI – Microcomputers (1970 – 1980).
5. Enhanced *VLSI*, *RISC*, *CISC* – *PCs*, *Supercomputers* (1980+).

The *Personal Computer*, or *PC*, has evolved dramatically from 1980 on. Computer systems are designed to assist individuals with tasks. They come in a variety of shapes, sizes, capabilities, performance and specialties. The computer system is an intricate arrangement of high technology components assembled together in a perfect way to make that machine assist the user. Since the *ENIAC* digital computer, this industry has been revolutionized in designing the mainframes where limited corporations owned, to minicomputers where large and mid size corporations could purchase, to the microcomputers for the hobbyist and the curious to the introduction of the Personal Computer in the early 1980's for all the corporations and small businesses as well as homes and schools. Computer systems range from a relatively simple "dumb" terminal to a large and a very complex mainframe or supercomputer. Computer systems are simple in performing any task. The complex task is broken into one or more simple subtasks and so on. With the continuous change in computer system technology and performance, it is becoming confusing to understand the computer architecture.

Depending on the limitation of its electronics circuits, a typical computer can perform these three basic functions:

a) *Arithmetic Operations* (add, subtract, divide and multiply).
b) *Logical Operations* (greater than, equal to and less than).
c) *Storage and Retrieval Operations* (read/write hard disk).

The primary reason that computers are being used more and more at home and at work is because:

    a) *Speed*: Modern computers can perform millions of calculations in one second.

    b) *Accuracy*: The accuracy of computers is depended on the accuracy of its input. This accuracy refers to the inherent reliability of its electronics circuits. The computer can run for hours, days and weeks at a time, outputting the expected results. If the input data is faulty or scrambled, then the output is useless and meaningless.

    c) *Storage*: Computers can store large amounts of information and data in a small device. Information and files that used to be stored in four drawers file cabinets (over 15,000 cubic inches) can easily be stored on a floppy disk (9 cubic inches).

    d) *Memory:* Computers can remember any size of task based pre-programmed steps.

This book attempts to:

1. Give a brief history on computers.
2. Define the computer system (how is it put together).
3. Describe hardware components (how do they function).
4. Explain Operating Systems (what is it).
5. Computer Virus (how does it work, how to stop it).
6. Help select a computer (how to make a choice).

CHAPTER I
PC HISTORY

In 1955, *IBM* introduced the "*First Generation*" computer systems with the *model 704*. The *IBM 704* mainframe computer used parallel binary arithmetic circuits and other advanced mathematical calculation circuitry. This computer had advanced arithmetic operations but the Input/Output operations were slow which caused a bottleneck in the computer system. To resolve this problem, *IBM* introduced *model 709* which had *I/O Processors* (*or Channels*).

The inefficiency of vacuum tubes and the demand for faster and better computers challenged the scientists and the industry at that time to improve the electrical and electronic circuitries; hence, the birth of the computer industry. Vacuum tubes were then replaced with electronic transistors primarily for reliability performance. Then later, the combination of transistors, resistors, capacitors and diodes were replaced with Integrated Circuits (IC).

The primary characteristics of the *First Generation Computers* are:

- Vacuum tubes for internal operations.
- Magnetic drums for limited internal storage.
- Heat, maintenance and support problems.
- Punched cards for input and output.
- Slow system for Input, Process and Output.
- Low-level (machine code) programming languages.

From 1958 to 1964, the "*Second Generation*" of computer systems was introduced. This generation of mainframe computers was developed based on transistor technology. The transistor was invented in 1947 but was not applied to computers until this period. *IBM* re-engineered its *709 model* to use the transistor technology and renamed the *IBM 7090*. This mainframe computer was capable of calculating close to 500,000 additions per second.

The primary characteristics of the *Second Generation Computers* are:

- *Transistors* for internal operations.
- Increased magnetic cores for internal storage.
- External magnetic storage (disks and tapes).
- Reduction in heat dissipation and size of system.
- Increased processing speed and system reliability.
- Increase in using high-level programming language.

The "*Third Generation*" of computer systems was developed based on *Integrated Circuit* (*IC*) technology. In the late 1960's and early 1970's, Mainframe computers were popular for large organizations but not easily accessible or affordable for small businesses and residential homes. These mainframe computers were designed primarily with *Printed Circuit Boards* (*PCB*) digital circuitry to process information. Integrated circuits were just beginning to replace the large *PCB*. The CPU that today is smaller than a dime (US currency) would sometimes occupy multiple *PCB*s of about or over 1 square feet in size. Computer time, i.e., usage, support and maintenance, was very expensive and most of the time required special technical skills. During this time, all residential homes and most small and mid size businesses were not capable of purchasing or using such a system.

The primary characteristics of the *Third Generation Computers* are:

- *Integrated Circuits (IC)* for internal operations.

1

- Increased internal storage capacity.
- Compatible systems (new to old).
- Introduction of *Minicomputers*.
- Reduction in cost and size of system.
- Increase in processing speed and system reliability.
- Increase high-level programming language industry.
- Operating Systems on external media storage.

In 1981, *IBM* introduced the *IBM Personal Computer* or *PC*. The brand name *"PC"* was then so increasingly used that to this date, it has come to mean an *IBM* compatible *Personal Computer*, not a *Macintosh*, a *VAX*, a *UNIX* or any other computer system. Within a few years, *IBM* introduced an enhanced PC, *PC-XT* that could be expanded easily, then an advanced PC, *PC-AT*. Both PCs were compatible with the original PC. The term *compatible*, indicating *PC compatible* was widely used as an industry standard. It described the ability to use one's manufacturer's component (hardware or software) with another manufacturer's component on the same PC. It also referred to the computer systems of a manufacturer that are like other manufacturer's computers, e.g., *Compaq* computers are compatible to *IBM* computers. A computer that can replace the hardware or the software but still run *DOS*, *Windows* and the so-like applications of a *PC* is a truly *IBM PC compatible*.

The primary characteristics of the *Fourth Generation Computers* are:

- *Large-Scale Integration (LSI)* for internal operations.
- Introduction of the Microprocessor and *Reduced Instruction Set Computer (RISC)* computers.
- Introduction of the *Micro* and *Super Computers*.
- Flexible and versatile software industry.
- Introduction of the *Very-Large-Scale Integration (VLSI)*.
- Increased internal storage capacity.
- Compatible systems (new to old).
- Reduction in cost and size of system.
- Increased processing speed and system reliability.
- Introduction of color printers, plotters (*Microcomputers*).

The *"Fifth Generation"* of computer systems refers to computers that were introduced after the year 1991. This generation of computers is known for high-performance (speed, uptime and reliability) capability and low-cost hardware. Some of these computers use multiple Processors, i.e., Processors are parallel to each other, operate simultaneously, may process tasks or sub-tasks independently from each other. These machines are referred to as *Parallel Computers* (or *Supercomputers*). They have a substantial increase in overall computer power (speed, response time, etc.) compared to a uni-processor (single Processor) computer. They are capable of executing tens of billions of CPU instructions per second. Examples of *Supercomputers* are the *Cray*, *NEC* and *Fujitsu* systems.

The *Complex Instruction Set Computer (CISC)* computer was also developed in this generation. The *CISC* machine has complex CPU instruction set that can perform complex operations. It has less CPU instructions in the program but the overall computer performance (speed especially) is better than the typical Personal Computer. The difference between *CISC* and *RISC* is that the *CISC* has a very complex CPU instruction set to perform complex operations while the *RISC* (developed in the *Fourth Generation*) employs a simpler CPU instruction set to improve the rate at which instructions can be executed. The less CPU instructions required, the shorter the program, the less time to execute a program, the faster the execution and completion of a task. The *CISC* emphasizes reducing the number of CPU instructions in the program, hence, increasing the overall performance of the computer. The *RISC* emphasizes reducing the average number of clock cycles required to execute an instruction, rather than reducing the number of instructions in the program. In *RISC*, most CPU instructions are executed within a single cycle. The *Intel Pentium* CPU is *CISC* architecture. The *IBM/Apple/Motorola PowerPC* CPU series, *Motorola* 88000 CPU series and the *DEC Alpha AXP* CPU are *RISC* architectures.

The primary characteristics of the *Fifth Generation Computers* are:

- Introduction of the *CISC* Microprocessor.
- Flexible and versatile software industry.
- Improved Very-Large-Scale Integration (*VLSI*).
- Increased internal and external storage capacity.
- Compatible systems (new to old).
- Introduction of Palm computers.
- Reduction in cost and size of system.
- Increased processing speed and system reliability.
- Introduction of scanners, digital cameras, etc...

In general, computers of all generations are alike. The technology has changed over the years but performance was improved. The illustration of this typical model may explain how computers function. The **Central Processing Unit** and the **Memory Storage Unit** are the **Process** of this model (*Figure 1.0*).

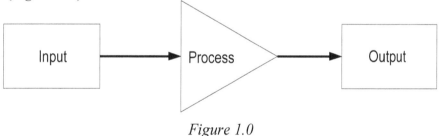

Figure 1.0

The hardware of a computer system basically consists of four functional sections, as shown in the following block diagram (*Figure 1.1*): The **Input Unit** (keyboard, mouse, ROM, RAM, floppy disk, hard disk, etc...), the **Central Processing Unit** (80486, *Pentium*, etc...), the **Memory Storage Unit** (ROM, RAM, floppy disk, hard disk, etc...) and the **Output Unit** (video monitor, printer, RAM, floppy disk, hard disk etc...). Each of these sections has a special function in terms of the overall computer operation.

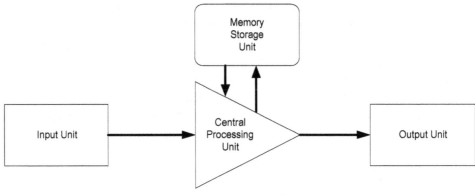

Figure 1.1

As a summary, Computer Systems are in four categories:

    a. *Microcomputer*: A single user PC with limited CPU speed, memory and storage capacity. Example: a standalone desktop or a laptop. This is a flexible computer because:

        a. Can run many applications as a standalone.
        b. Can be attached to a network of computers.
        c. Easy, low-cost hardware and software upgrade.

    b. *Minicomputer*: A multi-user PC with high speed CPU, large memory and storage capacity. Example: *Unix* or *NT* server. This type of computer is a little bit more powerful than and almost as flexible as the Microcomputer but has more system security built-in.

    c. *Mainframe computer*: A physically large size computer with considerable high speed CPU and a very large size memory and storage capacity. A special secured computer room with controlled cooling system (air conditioned) is normally required for this type of computer. Also, special electrical wiring with noise and electric surge suppression devices are usually installed.

    d. *Supercomputer*: A large-scale computer system (not necessarily physical size) designed for complicated tasks (require complex and lengthy calculations) and scientific applications such as weather forecasting, universities research centers, aerospace, etc... In most cases, these computers have a Parallel Processors. They are usually the largest in devices (memory, CPU, etc...) and fastest (processing). Example: *Cray Research, NEC* and *Fujitsu*.

CHAPTER II
BASIC ELECTRONIC SYSTEM

Computers are constructed of electronic circuitry: mostly *resistors, capacitors, transistors* and *Integrated Circuits (IC)*. The output of most electronic circuits is an *analog quantity* such as a voltage. An *analog quantity* is a quantity that may assume any numerical value within the range of possible output or outputs, it is a continuously variable quantity; one that may assume any value within a limited range. Therefore, an electronic circuit is capable of producing one or many outputs in response to different inputs. Analog signals (inputs or outputs) may represent the human voice (audio), light variations (video) or other physical phenomena.

Digital circuits are a part of the analog circuits with the exception that the output can be one of two values, called a ***bit***: a logical one (1) bit or a logical zero (0) bit. Digital signals are usually pulses, either *ON* or *OFF*. These pulses or voltage levels are manipulated to produce results in terms of discrete quantities, such as numbers. In troubleshooting a digital circuit, the output of a circuit, a single bit, is now minimized to the possible answer of either a logic zero or a one. Hence reducing the confusion of the multiple values that would occur in a normal analog circuit such as the value of *voltage* (*Volts*), AC or DC, the value of *current* (*Amps*), etc…

- *Basic electrical definitions and formulas*

Electricity: The electricity is a property of electrons and protons, expressed numerically in coulombs.

Electronic: *Electronic* is using or activated by electric current in semiconductors such as transistors, capacitors, and diodes.

Electronics: *Electronics* is the field of technology that deals with electronic devices such as transistors, capacitors and diodes.

Ampere *(André Marie Ampère, 1775-1836):* The current (I) that would produce a force of 2 * 10^{-7} *Newton* per meter of length between two parallel conductors one meter apart. It is equal to a charge flow of 1 coulomb per second. The symbol is *A*. 1 *ampere* = 1 *coulomb/second*.

Coulomb *(Charles Augustin de Coulomb, 1736-1806):* The charge (Q) that passes by a point in one second when the current is one ampere. It is equal to the charge of 6.242 * 10^{18} electrons. The symbol is *C*.

Watt *(James Watt, 1736-1819):* The power (P) required to do work at a rate of one joule per second. The symbol is *W*. 1 *Watt*=1 *volt* * 1*ampere*.

Joule *(James Prescott Joule, 1818-1889):* The work (W) done by a force of one Newton acting through a distance of one meter. The symbol is *J*. *Joule* is the unit of energy,

$$1 \ Joule = \frac{1 kilogram - meter^2}{second^2} \quad (2.0)$$

Volt *(Alessandro Volta, 1745-1827):* The potential difference between two points on a wire carrying one ampere of current when the power dissipated between the points is one watt. The symbol is *V*. 1*Volt*=1 *Joule /coulomb*.

Ohm *(George Simon Ohm, 1787-1854):* The resistance (Ω) that produces a voltage of one volt when carrying a current of one ampere. The symbol is **Ω**. 1 *ohm* = $\dfrac{1 volt}{1 ampere}$ *(2.1)*

***Farad** (Michael Faraday, 1791-1867):* The capacitance (C) in which a charge of one coulomb produces a potential difference of one volt. The symbol is **F**. 1 *Farad* = 1 *coulomb / volt.*

- *The Unit abbreviations*

Unit Symbol	Unit Prefix	Greek Origin	Decimal Power	Decimal measurements	Binary Power	Binary measurements	Meaning
T	Tera	Teras	10^{12}	1,000,000,000,000	2^{40}	1,099,511,627,776	trillion
G	Giga	Gigas	10^{9}	1,000,000,000	2^{30}	1,073,741,824	billion
M	Mega	Megas	10^{6}	1,000,000	2^{20}	1,048,576	million
K	Kilo	Khilioi	10^{3}	1,000	2^{10}	1,024	thousand
h	Hecto		10^{2}	100			hundred
da	deka		10^{1}	10			ten
d	deci		10^{-1}	0.1			One tenth
c	centi		10^{-2}	0.01			One hundredth
m	milli		10^{-3}	0.001			One thousandth
μ	micro		10^{-6}	0.000,000,1			One millionth
n	nano		10^{-9}	0.000,000,000,1			One billionth
p	pico		10^{-12}	0.000,000,000,000,1			One trillionth

Table 2.0

Note:

A decimal Mega is 10^{6}; e.g., 1 MegaWatts is 1 MW = 1,000,000 Watts.

A Binary Mega is 2^{20} or 1,048,576; e.g., 1 Mbytes = 1,048,576 bytes.

- *Basic electronic components*

Resistor, (*Figure 2.0*), is used to resist (reduce) the flow of electricity (current) in a path by a certain amount. Resistance unit is Ohms (Ω).

The electrical symbol

resistor samples

resistor, capacitor and IC

Figure 2.0

Capacitor, (*Figure 2.1*), is used to store electrical charge for a period of time then releases it. Capacitor unit is Farad (F).

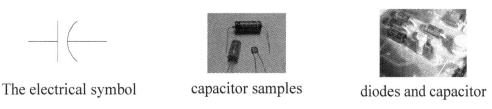

The electrical symbol   capacitor samples   diodes and capacitor

Figure 2.1

Diode, (*Figure 2.2*), is a natural switch because it operates in either of two logic states:
1. OFF state, its resistance is very high.
2. ON state, it passes current easily.

The electrical symbol   diodes

Figure 2.2

Transistor, (*Figure 2.3*), discovered in 1947 by a team of *Bell Telephone Laboratory* scientists. Before the transistor technology, relays and switches were the original digital logic devices but because of many reasons, among them high electrical power and make/break contact speed, they were replaced with the transistors (Relays required high electrical voltage and current - hence generated heat, they were slow in making and breaking contact). *Transistors* are used in a circuit as a switch, a small voltage is applied at one pole that switches a larger voltage at the other poles (ON or OFF). The transistor is the basic block from which all integrated circuits are built. The transistor can only create binary information: a logical 1 (voltage ON) if current passes through the poles or a logical 0 (voltage OFF) if current does not. The logics 1 and 0 individually are called **bits**.

The electrical symbol   transistors: samples and installed
(PNP = Positive-Negative- Positive)

Figure 2.3

Transistors can also be used to amplify an electrical signal. That is, they can increase the magnitude of the input signal current or voltage (or both) to a higher voltage current or power level by controlling a power supply voltage. The basic transistor switch circuit is illustrated in (*Figure 2.4*),

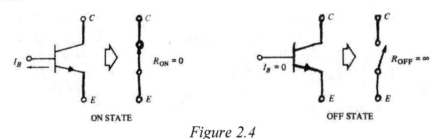

Figure 2.4

A **Logic gate** is a circuit with two or more inputs and whose output will be either a high or a low voltage. It is a logic element that passes or blocks an electrical signal. The Logic gates are widely used in digital computers and in all types of digital circuits and systems. Computers also use Boolean algebra, which is different than ordinary algebra. A Boolean algebra is a quantity that may, at different times, be equal to either logic 0 or logic 1. A Boolean logic 0 or 1 does not represent an actual number but instead represents a state of a voltage variable or logic level (*Table 2.1*). A voltage in a digital circuit is said to be at the logic level 0 or the logic level 1 depending on its actual numerical value. In a digital logic circuit, there are several terms that are used synonymously with logic 0 and logic 1, as shown here:

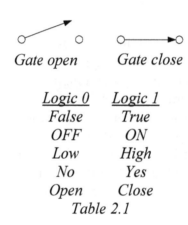

Gate open   *Gate close*

Logic 0	*Logic 1*
False	*True*
OFF	*ON*
Low	*High*
No	*Yes*
Open	*Close*

Table 2.1

There are five basic electronic **logic gates**:

1. **AND gate:** The *AND* gate logic output is equal to the *AND* product of the logic inputs, or,
   *Output x = Output (AB) = Input A * Input B (2.2)*
   As shown in the *Truth* table (*Table 2.2*),

A	*B*	*x = A * B*
0	*0*	*0*
0	*1*	*0*
1	*0*	*0*
1	*1*	*1*

Table 2.2

AND gate, (*Figure 2.5*), electrical circuit and symbol are:

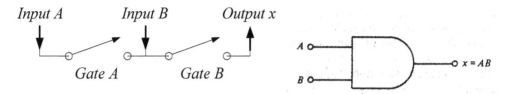

Figure 2.5

2. ***OR gate***: The *OR* gate logic output is equal to the *OR* sum of the logic inputs, or,
    Output x = Output (A+B) = Input A + Input B (2.3)
    As shown in the *Truth* table (*Table 2.3*),

A	B	x = A + B
0	0	0
0	1	1
1	0	1
1	1	1

Table 2.3

OR gate, (*Figure 2.6*), electrical circuit and symbol are:

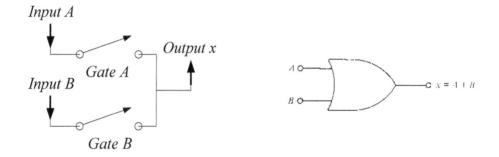

Figure 2.6

1. ***NOT gate***: The NOT gate (inverter) logic output is equal to the inverse (complement) of the logic input, or,
    Output x = Complement Input A (2.4)
    As shown in the *Truth* table (*Table 2.4*),

A	x = complement A
0	1
0	0

Table 2.4

NOT gate (inverter)***,*** (*Figure 2.7*), electrical circuit and symbol are:

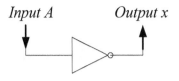

Input A   *Output x*

Figure 2.7

2. **NAND gate:** The NAND gate is equivalent to the AND gate followed by a NOT gate (an INVERTER).
Output x = Complement (Input A * Input B) (2.5)
As shown in the *Truth* table (*Table 2.5*),

A	B	x = Complement (A * B)
0	0	1
0	1	1
1	0	1
1	1	0

Table 2.5

NAND gate, (*Figure 2.8*), electrical circuit and symbol are:

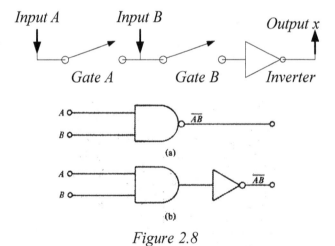

Input A   *Input B*   *Output x*

Gate A   *Gate B*   *Inverter*

Figure 2.8

3. **NOR gate:** The NOR gate is equivalent to the OR gate followed by a NOT gate (an INVERTER).
Output x = Complement (Input A + Input B) (2.6)
As shown in the *Truth* table (*Table 2.6*),

10

A	B	x = Complement (A + B)
0	0	1
0	1	0
1	0	0
1	1	0

Table 2.6

NOR gate, (*Figure 2.9*), electrical circuit and symbol are:

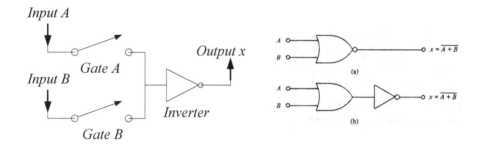

Figure 2.9

Integrated Circuits (IC), invented in 1957, is a tiny chip (smaller than a dime [US currency]) of silicon that hosts thousands or millions of transistors and other circuits components (resistors, capacitors, etc…). The silicon is encased in a sturdy ceramic (or other non-conductive) material. Small conductive wires that are extruded from the IC plug into the Printed Circuit Board of a computer, connecting the encased chip to the computer (conductivity). ICs (*Figure 2.10*) are faster, smaller, cheaper, and more reliable and consume less power than relays or vacuum tubes.

Figure 2.10

There are many types of ICs, (*Figure 2.11*): *Dual Inline Package (DIP), Quad Small Outline Package (QSOP* or *surface mount), Scale Integration (VLSI),* Very Large *Single Inline Package (SIP)* and *Pin Grid Array Package (PGA).*

There are many types of IC logic families, among them:

Resistor Transistor Logic (RTL): The *RTL* is built of resistors and transistors. It was the earliest of the logic family developed. An example of RTL is a NOR gate.

Diode Transistor Logic (DTL): The *DTL* was developed as a result of some of the problems in the RTL logic such as voltage variation and low fanout (number of loads that can be driven from a single source or input). An example of DTL is a NAND gate.

Transistor Transistor Logic (TTL)*:* The *TTL* displaced the *DTL* because of the speed and low impedance in both logic 1 and logic 0 output states. An example of *TTL* is a *NAND* gate.

Emitter Coupled Logic (ECL)*:* The *ECL* is faster than *TTL*. It is used mostly in circuits where high speed (propagation delay time) is essential. Examples of ECL are OR and NOR gates.

Complementary Metal Oxide Semiconductor Logic (CMOS)*:* The *CMOS* logic is popular because of low power dissipation, hence low heat. Example of *CMOS* is memory ICs.

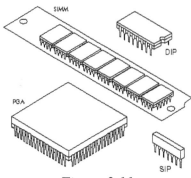

Figure 2.11

The many advances that have taken place in the field of electronics since the early 1970's have led to rapid advances in computer system technology. For instance, the introduction of the **Small-Scale Integrated (SSI)** circuits (1 to 10 transistors) followed by the **Medium-Scale Integrated (MSI)** circuits (10 to 100 transistors), then the **Large-Scale Integrated (LSI)** circuits (100 to 5,000 transistors), then the **Very-Large-Scale Integrated (VLSI)** circuits (5,000 to 50,000 transistors), the **Super-Large-Scale Integrated (SLSI)** circuits (50,000 to 100,000 transistors) and then **Ultra-Large-Scale Integrated (ULSI)** circuits (over 100,000 transistors).

These circuitries have led the way in expanding the performance and capacity of the Personal Computers (desktops and laptops). With thousands of transistors, a computer can create any number, provided that it has enough transistors grouped together to hold a logical 1 and logical 0 states. If *eight outputs* of transistors are grouped together as *eight bits* then this group is called a *byte*. A group of *two bytes* is called a *word*. A *word* is 16-bits and a *byte* is 8-bits.

A *chip* (*Figure 2.12*), is a small piece of a semiconductor material (typically silicon) on which an IC is embedded. A typical chip is less than ¼-square inches and can contain millions of electronic components (usually transistors). Computer Systems consist of many chips placed on electronic boards called *Printed Circuit Boards (PCB)*.

Figure 2.12

There are different types of chips. For example, the *CPU* chip (also called *Microprocessor* or *Processor*) contains an entire processing unit, whereas memory chip contain blank memory cells. Chips come in a variety of packages.

The three most common chips are:

- **Dual In-line Package (DIP):** The *DIP,* (*Figure 2.13*), is the traditional bug like chip that have anywhere from 8 to 40 legs, evenly divided in two rows.

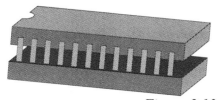 

Figure 2.13

- **Pin-Grid Array (PGA):** The *PGA*, (*Figure 2.14*), is a square chip in which the pins are arranged in concentric squares. The PGA chips are particularly good for chips that have many pins, such as the microprocessor (CPU) manufactured in the early 1990s and later. In the mid 1980's, *IBM* developed *Professional Graphics Adapter (PGA)*. This was a video standard that supported 640x480 resolution.

Figure 2.14

- **Single In-line Package (SIP):** The *SIP*, (*Figure 2.15*), is an IC that has just one row of legs in a straight line like a comb.

Figure 2.15

- **Single In-line Memory Module (SIMM):** The *SIMM*, (*Figure 2.16*), is a *PCB* that consists of nine or more packaged as a single unit.

Figure 2.16

Carlos E. Hattab

- *Printed Circuit Board (PCB)*

The **Printed Circuit Board (PCB)**, (*Figure 2.17*), is a thin plastic plate on which ICs, resistors, capacitors and other electronic components are placed. Computer systems consist of one or more PCBs, some are called **cards** or **adapters**. Circuit boards fall into the following categories:

- **Motherboard (or mainboard or system board):** The principal board that has the most connectors (video, sound, network cards) for attaching devices to the Bus. Typically, the motherboard contains the CPU, Memory (ROM, RAM and CMOS), and other basic controllers (floppy, hard and CD-ROM drives) for the computer system.
- **Expansion board:** Any board that plugs into one of the computer's expansion slots. Expansion boards include I/O controller boards, video adapters, network cards, and others.
- **Daughter card:** Any board or card that attaches directly to another board from the expansion slot.
- **Controller board:** A special type of expansion board that contains a controller for a peripheral device (disk drive or graphics monitor).

 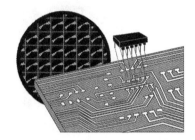

Figure 2.17

14

CHAPTER III
NUMBERING SYSTEM

Computer systems are often called "*number cruncher*" systems, because computers basically resolve all problems and questions, through logic, with numbers. They also display graphics images, playback audio sounds and scan pictures by converting and processing numbers.

- *Decimal*

The *Decimal* number is the basic numbering system for counting. The word decimal was derived from the *Latin* word "*decem*", meaning ten. The decimal number system is positional or weighted. That means each digit position in a number carries a particular weight, which determines the magnitude of that number. Each position has a weight determined by some power of the number base. In this case, the decimal base is 10; there are ten digits in this counting system: 0, 1,2,3,4,5,6,7,8 and 9. The positional weights are 10^0(units), 10^1(tens), 10^2(hundreds), etc. The *Least Significant Digit* or *LSD*, is the right-most digit of the number because it carries the lowest weight in determining the value of the number. The left-most digit is the *Most Significant Digit* or *MSD*, because it carries the greatest weight in determining the value of the number. For example, the number 34,215 has a 3 as MSD and 5 as LSD. Computers do not use *Decimal* base system, instead, they use *Binary* and *Hexadecimal* base systems.

Note: Any number with an exponent of zero equals to 1, e.g., $100^0 = 1$.

- *Binary*

The *Binary* is the simplest number system that uses positional notation. The binary is base 2, it contains only two elements or states; therefore the digits can only be a zero (0) or a one (1). A binary number is a group of zeros (0s) and ones (1s). It is the logical language that the CPU of a computer can understand. The number 2 does not exist in the binary system, just as there is no number beyond 9 in the decimal system. The CPU has to interpret (through its transistors - switches) all the inputs and make the decision through outputs. Hence, the Binary base numbers are the native tongue or language of computers. An open switch is a logical 0 (zero) and a closed switch is a logical 1 (one).

As thousands of instructions are part of the *Basic Input/Output System (BIOS)* or *Operating System*, the language that the CPU interprets, is known as the *Machine Language*. An *Assembly Language* is the lowest computer language. The *Assembly Language* uses mnemonic names to stand for one or more *Machine language* instructions. *Machine language* is the most basic software instruction in the computer, and *Assembly Language* is a "shorthand" method of representing the long strings of ones and zeros found in *Machine language*. The advantage of *Assembly Language* over high-level languages (such as *BASIC*, *COBOL* and *PASCAL*) is its speed for applications that require frequent and quick data transfer. High-level languages are generally easier to use than assembly language. Example of a *500 μsec.* delay count using an Intel 8085 microprocessor:

Memory Address	Machine Language	Assembly Language	Program comment
F800	06 6B	MVI B, 6B	;move 6B hex into register B
F802	05	DCR B	;decrement reg. B value (6B H)
F803	C2 02F8	JNZ F802	;jump loop until value = 0

(**Note:** *RAM* addressing begins at F800 hex. The 8085 CPU takes 4.66 µsec. each loop. For the 500 µsec. delay count,

$$\text{Count} = \frac{500}{4.66} = 107 \text{ decimal or 6B hex.}$$

A single binary digit is called a *bit*, such as 0 or 1. The term is derived from the contraction of **b**inary dig**it**. A group of two bits is called *dibit*, or *Slice* such as 01. A group of four bits associated together are called a *nibble*, such as (0000). A nibble can also be referred to as *Binary Coded Decimal*, or *BCD*. A group of eight bits or two nibbles associated together are called a *byte*, such as (1011 1100). The right-most bit in a binary number is the *Least Significant Bit* or *LSB*. The left-most bit is the *Most Significant Bit*, or *MSB*. For example, the binary number 10111100 has 0 as *LSB* and 1 as *MSB*. The *MSD* nibble of a byte is called the *high nibble* and the *LSD* nibble of a byte is called the *low nibble*. For example, the number 1011 1100 has 1011 as *high nibble* and 1100 as *low nibble*. A word is a group of bits that constitutes the basic unit of information within a computer system. In a 16-bit system, a group of 16 bits, or four nibbles or two bytes associated together are called a *word*, such as (0101 1111 0000 1010). A word in a 32-bit system is four bytes. The *MSD* byte of a word is called the *high byte* and the *LSD* byte of a word is called the *low byte*. For example, the number 0101 1111 0000 1010 has 0101 1111 as *high byte* and 0000 1010 as *low byte*. In a 16-bit system, a group of 32 bits, associated together is called *two words*. A group of 64 bits, associated together is called *four words* and so on.

Now, putting all these terms together, the general representation of a bit is normally as a nibble, a byte or a word, example, a decimal 1 is a binary 0001. The conversion of two 4-bit digits: 0001 and 1011 from binary to decimal is as follows, (Note: The factor is a multiplier of 2, 2^3 is read as 2 to the power of 3, where 3 denotes the 4th bit of the number –since the first bit begins from 0 (zero)).

$$\begin{array}{cccc} 2^3 & 2^2 & 2^1 & 2^0 \\ 0 & 0 & 0 & 1_2 \end{array}$$

From algebra, $2^0 = 1$,

$$(0 * 2^3) + (0 * 2^2) + (0 * 2^1) + (1 * 2^0)$$
$$(0) + (0) + (0) + (1) = 1_{10}$$

hence 0001_2 (binary) is 1_{10} (decimal).

$$\begin{array}{cccc} 2^3 & 2^2 & 2^1 & 2^0 \\ 1 & 0 & 1 & 1_2 \end{array}$$

From algebra, $2^0 = 1$, $2^1 = 2$ and $2^3 = 8$

$$(1 * 2^3) + (0 * 2^2) + (1 * 2^1) + (1 * 2^0)$$
$$(8) + (0) + (2) + (1) = 11_{10}$$

hence 1011_2 (binary) is 11_{10} (decimal).

Converting a 16-bit digit such as $0101\ 1111\ 0000\ 1010_2$ requires more details, for example,

2^{15}	2^{14}	2^{13}	2^{12}	2^{11}	2^{10}	2^9	2^8	2^7	2^6	2^5	2^4	2^3	2^2	2^1	2^0
0	1	0	1	1	1	1	1	0	0	0	0	1	0	1	0

$(0 * 2^{15}) + (1 * 2^{14}) + (0 * 2^{13}) + (1 * 2^{12}) + (1 * 2^{11}) + (1 * 2^{10}) +$
$(1 * 2^9) + (1 * 2^8) + (0 * 2^7) + (0 * 2^6) + (0 * 2^5) + (0 * 2^4) + (1 * 2^3) +$
$(0 * 2^2) + (1 * 2^1) + (0 * 2^0)$
$(0) + (16,384) + (0) + (4,096) + (2,048) + (1,024) + (512) + (256) + (0) + (0) + (0) + (0) + (8) +$
$(0) + (2) + (0) = 24,330_{10}$ (decimal)

Note that bit 16 is 2 to the power of 15, that is because the first bit always begins at 0 zero, i.e., bit 64 is 2 to the power of 63.

Binary, in computer world, is bit 1 or 0. Therefore, a 2-bit is 2^2 or 4 lines. That means a CPU can handle 4 lines. Other CPU examples:

2-bit is $2^2 =$	4
4-bit is 2^4 or 2*2*2*2 =	16
8-bit is 2^8 or 2*2*2*2*2*2*2*2 =	256
16-bit is 2^{16} or	65,536
32-bit is 2^{32} or	4,294,967,295 (4 GigaBytes, GB)
64-bit is 2^{64} or	18,446,744,073,709,551,616

In computer count,

1 Kilobits	$= 2^{10} =$	1,024 bits
1 Megabits	$= 2^{20} =$	1,048,576 bits
1 Gigabits	$= 2^{30} =$	1,073,741,824 bits

To isolate the decimal from binary measures, the *Institute of Electrical and Electronics Engineers (IEEE)* has proposed a new naming convention for the binary numbers, to hopefully eliminate some of the confusion. Under this proposal, for binary numbers the third and fourth letters in the prefix are changed to "*bi*", so "*Mega*" becomes "*Mebi*" for example. Thus, one Megabyte would be 10^6 bytes, but one *Mebibyte* would be 2^{20} bytes. The abbreviation would become "1 *MiB*" instead of "1 *MB*".

- *Octal*

The *Octal* is another number system that is often used with microprocessors. It has a base 8 numbering system, and uses the digits 0,1,2,3,4,5,6,7. As with the binary number system, each digit position of an octal number carries a positional weight, which determines the magnitude of that number. The weight of each position is determined by some power of the number system base, in this case, it is 8. For example, the number 302_8 (octal) is,

$$(3 * 8^2) + (0 * 8^1) + (2 * 8^0) = (3 * 64) + (0) + (2 * 1) = 194_{10}$$

hence, 302_8 (octal) is 194_{10} (decimal).

The octal base was used in computer systems during the 1970's. For example, some of the *Digital Equipment Corporation's PDP* computers used octal numbering system for processing information.

Microprocessors of computers manipulate data using the binary number system. However, when larger quantities are involved, the binary number system can become cumbersome. Therefore, other number systems are frequently used as a form of binary shorthand to speed up and simplify data entry and display. The octal and hexadecimal are the numbering systems used. Both numbering systems are similar to the decimal number system, which makes it easier to understand numerical value. In addition, conversion between binary and octal (or hexadecimal) is readily accomplished because of the value structure of octal (or hexadecimal). For example, the binary $101\ 001_2$ is 51_8.

In the late 1970's, with the introduction of the *CP/M* microcomputer, binary and hex numbering systems replaced the octal.

- *Hexadecimal*

The *Hexadecimal* is another numbering system used often with microprocessors. It is similar in value structure to the octal number system, and thus allows easy conversion with the binary numbering system. Because of this feature and the fact that hexadecimal simplifies data entry and display to a greater degree than octal, there are more microprocessors in computers using the hexadecimal numbering system rather than the octal. As computers are processing and crunching numbers, it is inefficient and complicated to represent any large numbers in binary. It would be very easy and more efficient to use the *Hexadecimal* base numbering system.

The hexadecimal (also called hex) is base 16, it uses the decimal numbers ($0_{16} = 0_{10}$, $1_{16} = 1_{10}$, $2_{16} = 2_{10}$ through $9_{16} = 9_{10}$) and the letters A_{16} through F_{16} ($A_{16} = 10_{10}$, $B_{16} = 11_{10}$, $C_{16} = 12_{10}$, $D_{16} = 13_{10}$, $E_{16} = 14_{10}$, $F_{16} = 15_{10}$). For example, the number $E5D7_{16}$ is

$$(E * 16^3) + (5 * 16^2) + (D * 16^1) + (7 * 16^0)$$
$$(14 * 4096) + (5 * 256) + (13 * 16) + (7 * 1)$$
$$57,344 + 1,280 + 208 + 7 = 58,839_{10}$$

The number $24,330_{10}$ (decimal) is

$0101\ 1111\ 0000\ 1010_2$ (binary)

$\quad 5 \qquad F \qquad 0 \qquad\quad A_{16}$ (hex)

Table 3.0 is a conversion table of Decimal numbers from 1 to 255 to Binary, Hex and Octal.

- *American National Standards Institute (ANSI)*

The *ANSI* is the primary organization for fostering the development of technology standards in the United States. *ANSI* works with industry groups and is the U.S. member of the *International Organization for Standardization (ISO)* and the *International Electrotechnical Commission (IEC)*. Long-established computer standards from *ANSI* include the *American Standard Code for Information Interchange (ASCII)* and the *Small Computer System Interface (SCSI)*.

Carlos E. Hattab

				Numbering Base Table							
				RADIX							
Decimal	Binary	Hex	Octal	Decimal	Binary	Hex	Octal	Decimal	Binary	Hex	Octal
0	0000 0000	00	000	35	0010 0011	23	043	85	0101 0101	55	125
1	0000 0001	01	001	36	0010 0100	24	044	90	0101 1010	5A	132
2	0000 0010	02	002	37	0010 0101	25	045	95	0101 1111	5F	137
3	0000 0011	03	003	38	0010 0110	26	046	100	0110 0100	64	144
4	0000 0100	04	004	39	0010 0111	27	047	105	0110 1001	69	151
5	0000 0101	05	005	40	0010 1000	28	050	110	0110 1110	6E	156
6	0000 0110	06	006	41	0010 1001	29	051	115	0111 0011	73	163
7	0000 0111	07	007	42	0010 1010	2A	052	120	0111 1000	78	170
8	0000 1000	08	010	43	0010 1011	2B	053	125	0111 1101	7D	175
9	0000 1001	09	011	44	0010 1100	2C	054	130	1000 0010	82	202
10	0000 1010	0A	012	45	0010 1101	2D	055	135	1000 0111	87	207
11	0000 1011	0B	013	46	0010 1110	2E	056	140	1000 1100	8C	214
12	0000 1100	0C	014	47	0010 1111	2F	057	145	1001 0001	91	221
13	0000 1101	0D	015	48	0011 0000	30	060	150	1001 0110	96	226
14	0000 1110	0E	016	49	0011 0001	31	061	155	1001 1011	9B	233
15	0000 1111	0F	017	50	0011 0010	32	062	160	1010 0000	A0	240
16	0001 0000	10	020	51	0011 0011	33	063	165	1010 0101	A5	245
17	0001 0001	11	021	52	0011 0100	34	064	170	1010 1010	AA	252
18	0001 0010	12	022	53	0011 0101	35	065	175	1010 1111	AF	257
19	0001 0011	13	023	54	0011 0110	36	066	180	1011 0100	B4	264
20	0001 0100	14	024	55	0011 0111	37	067	185	1011 1001	B9	271
21	0001 0101	15	025	56	0011 1000	38	070	190	1011 1110	BE	276
22	0001 0110	16	026	57	0011 1001	39	071	195	1100 0011	C3	303
23	0001 0111	17	027	58	0011 1010	3A	072	200	1100 1000	C8	310
24	0001 1000	18	030	59	0011 1011	3B	073	205	1100 1101	CD	315
25	0001 1001	19	031	60	0011 1100	3C	074	210	1101 0010	D2	322
26	0001 1010	1A	032	61	0011 1101	3D	075	215	1101 0111	D7	327
27	0001 1011	1B	033	62	0011 1110	3E	076	220	1101 1100	DC	334
28	0001 1100	1C	034	63	0011 1111	3F	077	225	1110 0001	E1	341
29	0001 1101	1D	035	64	0100 0000	40	100	230	1110 0110	E6	346
30	0001 1110	1E	036	65	0100 0001	41	101	235	1110 1011	EB	353
31	0001 1111	1F	037	66	0100 0010	42	102	240	1111 0000	F0	360
32	0010 0000	20	040	70	0100 0110	46	106	245	1111 0101	F5	365
33	0010 0001	21	041	75	0100 1011	4B	113	250	1111 1010	FA	372
34	0010 0010	22	042	80	0101 0000	50	120	255	1111 1111	FF	377

Table 3.0

- *American Standard Code for Information Interchange (ASCII)*

The *ASCII* code is a system of computer coding for alphanumeric and punctuation characters, each character is assigned a number from 0 to 127 (*Table 3.1*). *ASCII* is the most common format for text files in computers and on the Internet. It is a 7-bit coding system used for representing characters in the computer. All these characters are given values (seven or eight binary digits). The table consists of 256 possible values. The first 32 values are control characters; they are assigned to computer communication and printer information. The remaining 96 values are numbers, alphabetic letters and common punctuation. Most computer systems use *ASCII* codes to represent text data information, which makes it possible to transfer data from one computer to another. In an *ASCII* file, each alphabetic, numeric, or special character is represented with a 7-bit binary number (a string of seven 0s or 1s). 128 possible characters are defined. Text files stored in *ASCII* format are sometimes called *ASCII* files. Text editors and word processors programs are usually capable of storing data in *ASCII* format, although ASCII format is not always the default storage format.

Most data files, particularly if they contain numeric data, are not stored in *ASCII* format. The standard *ASCII* character set uses just 7 bits for each character. There are several larger character sets that use 8 bits, which gives them 128 additional characters. The extra characters are used to represent non-English characters, graphics symbols, and mathematical symbols. Several companies and organizations have proposed extensions for these 128 characters. The DOS operating system uses a superset of *ASCII* called *extended ASCII* or *high ASCII*. UNIX and DOS-based operating systems (except for Windows NT) use *ASCII* for text files. Windows NT uses a newer code, Unicode.

ASCII Control Characters Definition:

Note: Caret A or ^A = Control A, i.e., hold the Ctrl key [key located on the keyboard of a PC] down and press the A key once. A control character (such as ^A) is a non-printing character.

NUL (Null or ^@): No character. Used to take up space or increase the time for input or output.

SOH (Start of heading or ^A): Contains addressing or other identification for a following message.

STX (Start of text or ^B): Terminates a heading and indicates a following message text.

ETX (End of text or ^C): Terminates message text.

EOT: (End of transmission or ^D): Terminates communications.

ENQ (Enquiry or ^E): Requests a response such as status or station identification.

ACK (Acknowledge or ^F): Affirmative response from receiver to transmitter.

BEL (Bell or ^G): Audible sound or indicator to alert receiver.

BS (Backspace or ^H): Moves cursor back one space on the same line.

HT (Horizontal tab or ^I): Moves cursor to the next tab on line.

LF (Line feed or ^J): Moves cursor down one line with no horizontal displacement.

VT (Vertical tab or ^K): Moves cursor down a preset number of lines.

FF (Form feed or ^L): Moves cursor to a preset line of the next page.

CR (Carriage return or ^M): Moves cursor to first character position of that line.

SO (Shift out or ^N): Used to extend the ASCII code to allow an extended graphics character set.

SI (Shift in or ^O): Terminates SO.

DLE: (Data link escape or ^P): Changes the meaning of a specific number of contiguously following characters.

DC1 (Device control 1 = X-ON or ^Q): Controls ancillary device or special feature.

DC2 (Device control 2 = Tape or ^R): Controls ancillary device or special feature.

DC3 (Device control 3 = X-OFF or ^S): Controls ancillary device or special feature.

DC4 (Device control 4 = Stop or ^T): Controls ancillary device or special feature.

NAK (Negative acknowledge or ^U): Negative response to transmitter from receiver.

SYN (Synchronous idle or ^V): Used to provide or achieve synchronism in data transmission.

ETB (End of transmission block or ^W): Used to indicate end of a data block.

CAN (Cancel or ^X): Used to indicate the preceding data is invalid.

EM (End of medium or ^Y): Used to identify the physical, used or wanted data recorded on a medium.

SUB (Substitute or ^Z): May replace an erroneous or invalid character.

ESC (Escape or ^[): Provides supplementary characters by changing the interpretation of the following:

FS (or ^\): File separator.

GS (or ^]): Group separator.

RS (or ^^): Record separator.

US (or ^_): Unit separator.

SP (Space): Moves cursor one space (blank character) forward without displaying or printing any character.

DEL (Delete): Used to obliterate or erase undesired characters.

Example from the *ASCII Table 3.1*: The letter *A* has a value of 41 *Hexadecimal*, 65 *Decimal* and 101 *Octal*. As a *byte*, it is: 0100 0001 and as a *word*, it is: 0000 0000 0100 0001

ASCII Code Chart															
Control				Numbers/Symbols				Upper case				Lower case			
Radix			ASCII	Radix			ASCII	Radix			ASCII	Radix			ASCII
Hex	Dec	Oct	Char	Hex	Dec	Oct	Char	Hex	Dec	Oct	Char	Hex	Dec	Oct	Char
00	0	000	NUL	20	32	040	SP	40	64	100	@	60	96	140	`
01	1	001	SOH	21	33	041	!	41	65	101	A	61	97	141	a
02	2	002	STX	22	34	042	"	42	66	102	B	62	98	142	b
03	3	003	ETX	23	35	043	#	43	67	103	C	63	99	143	c
04	4	004	EOT	24	36	044	$	44	68	104	D	64	100	144	d
05	5	005	ENQ	25	37	045	%	45	69	105	E	65	101	145	e
06	6	006	ACK	26	38	046	&	46	70	106	F	66	102	146	f
07	7	007	BEL	27	39	047	'	47	71	107	G	67	103	147	g
08	8	010	BS	28	40	050	(	48	72	110	H	68	104	150	h
09	9	011	HT	29	41	051	)	49	73	111	I	69	105	151	I
0A	10	012	LF	2A	42	052	*	4A	74	112	J	6A	106	152	j
0B	11	013	VT	2B	43	053	+	4B	75	113	K	6B	107	153	k
0C	12	014	FF	2C	44	054	,	4C	76	114	L	6C	108	154	l
0D	13	015	CR	2D	45	055	-	4D	77	115	M	6D	109	155	m
0E	14	016	SO	2E	46	056	.	4E	78	116	N	6E	110	156	n
0F	15	017	SI	2F	47	057	/	4F	79	117	O	6F	111	157	o
10	16	020	DLE	30	48	060	0	50	80	120	P	70	112	160	p
11	17	021	DC1	31	49	061	1	51	81	121	Q	71	113	161	q
12	18	022	DC2	32	50	062	2	52	82	122	R	72	114	162	r
13	19	023	DC3	33	51	063	3	53	83	123	S	73	115	163	s
14	20	024	DC4	34	52	064	4	54	84	124	T	74	116	164	t
15	21	025	NAK	35	53	065	5	55	85	125	U	75	117	165	u
16	22	026	SYN	36	54	066	6	56	86	126	V	76	118	166	v
17	23	027	ETB	37	55	067	7	57	87	127	W	77	119	167	w
18	24	030	CAN	38	56	070	8	58	88	130	X	78	120	170	x
19	25	031	EM	39	57	071	9	59	89	131	Y	79	121	171	y
1A	26	032	SUB	3A	58	072	:	5A	90	132	Z	7A	122	172	z
1B	27	033	ESC	3B	59	073	;	5B	91	133	[	7B	123	173	{
1C	28	034	FS	3C	60	074	<	5C	92	134	\	7C	124	174	|
1D	29	035	GS	3D	61	075	=	5D	93	135	]	7D	125	175	}
1E	30	036	RS	3E	62	076	>	5E	94	136	^	7E	126	176	~
1F	31	037	US	3F	63	077	?	5F	95	137	_	7F	127	177	DEL

Table 3.1

- *Unicode*

Unicode is an entirely new idea in setting up binary codes for text or script characters. Officially called the Unicode Worldwide Character Standard, it is a system for "the interchange, processing, and display of the written texts of the diverse languages of the modern world". It also supports many classical and historical texts in a number of languages. Currently, the Unicode standard contains 34,168 distinct coded characters derived from 24 supported language scripts.

These characters cover the principal written languages of the world. Additional work is underway to add the few modern languages not yet included. Also see the currently most prevalent script or text codes, *ASCII* and *EBCDIC*.

- *International Organization for Standardization (ISO)*

ISO, founded in 1946, is a worldwide federation of national standards bodies from some 100 countries, one from each country. Among the standards it fosters is *Open Systems Interconnection (OSI)*, a universal reference model for communication protocols. Many countries have national standards organizations such as the *American National Standards Institute (ANSI)* that participate in and contribute to *ISO* standards making. "*ISO*" is not an acronym. It is a word, derived from the Greek *isos*, meaning "*equal*", which is the root for the prefix "iso-" that occurs in a host of terms, such as "isometric" (of equal measure or dimensions) and "isonomy" (equality of laws, or of people before the law). The name *ISO* is used around the world to denote the organization, thus avoiding the plethora of abbreviations that would result from the translation of "*International Organization for Standardization*" into the different national languages of members, (for example, *IOS* in English or *OIN* in French - for *Organisation internationale de normalisation*). Whatever the country, the short form of the organization's name is always *ISO*.

- *Extended Binary Coded Decimal Interchange Code (EBCDIC)*

EBCDIC is a binary code for alphabetic and numeric characters that *IBM* developed for its larger operating systems. It is the code for text files that is used in *IBM's OS/390* operating system for its *S/390* servers and that thousands of corporations use for their legacy applications and databases. In an *EBCDIC* file, each alphabetic or numeric character is represented with an 8-bit binary number (a string of eight 0's or 1's). 256 possible characters (letters of the alphabet, numerals, and special characters) are defined. *IBM's PC* and workstation operating systems do not use *IBM's* proprietary *EBCDIC*. Instead, they use the industry standard code for text, *ASCII*. Conversion programs allow different operating systems to change a file from one code to another. The *EBCDIC* is a character set code (*Table 3.2 & Table 3.3*) that was used primarily on IBM mainframes for representing characters as numbers. It exists in at least six mutually incompatible versions and it is not compatible with the *ASCII* code. *IBM* adapted *EBCDIC* from the punched card code in the early 1960s. Example from the *EBCDIC Table 3.3*: The letter *A* has a value of C1 *Hexadecimal*, 193 *Decimal* and 301 *Octal*. As a *byte*, it is: 1100 0001. and as a *word*, it is: 0000 0000 1100 0001.

EBCDIC Code Chart (1 of 2)

Control				Numbers/Symbols				Upper case				Lower case			
Radix			EBCDIC	Radix			EBCDIC	Radix			EBCDIC	Radix			EBCDIC
Hex	Dec	Oct	Char	Hex	Dec	Oct	Char	Hex	Dec	Oct	Char	Hex	Dec	Oct	Char
00	0	000	NUL	20	32	040		40	64	100	SP	60	96	140	-
01	1	001		21	33	041		41	65	101		61	97	141	/
02	2	002		22	34	042		42	66	102		62	98	142	
03	3	003		23	35	043		43	67	103		63	99	143	
04	4	004		24	36	044		44	68	104		64	100	144	
05	5	005	PT	25	37	045		45	69	105		65	101	145	
06	6	006		26	38	046		46	70	106		66	102	146	
07	7	007		27	39	047		47	71	107		67	103	147	
08	8	010	GE	28	40	050	SA	48	72	110		68	104	150	
09	9	011		29	41	051	SFE	49	73	111		69	105	151	
0A	10	012		2A	42	052		4A	74	112		6A	106	152	
0B	11	013		2B	43	053		4B	75	113	.	6B	107	153	,
0C	12	014	FF	2C	44	054	MF	4C	76	114	<	6C	108	154	%
0D	13	015	CR	2D	45	055		4D	77	115	(	6D	109	155	_
0E	14	016		2E	46	056		4E	78	116	+	6E	110	156	>
0F	15	017		2F	47	057		4F	79	117	|	6F	111	157	?
10	16	020		30	48	060		50	80	120	&	70	112	160	
11	17	021	SBA	31	49	061		51	81	121		71	113	161	
12	18	022	EUA	32	50	062		52	82	122		72	114	162	
13	19	023	IC	33	51	063		53	83	123		73	115	163	
14	20	024		34	52	064		54	84	124		74	116	164	
15	21	025	NL	35	53	065		55	85	125		75	117	165	
16	22	026		36	54	066		56	86	126		76	118	166	
17	23	027		37	55	067		57	87	127		77	119	167	
18	24	030		38	56	070		58	88	130		78	120	170	
19	25	031	EM	39	57	071		59	89	131		79	121	171	'
1A	26	032		3A	58	072		5A	90	132	!	7A	122	172	:
1B	27	033		3B	59	073		5B	91	133	$	7B	123	173	#
1C	28	034	DUP	3C	60	074	RA	5C	92	134	*	7C	124	174	@
1D	29	035	SF	3D	61	075		5D	93	135	)	7D	125	175	'
1E	30	036	FM	3E	62	076		5E	94	136	;	7E	126	176	=
1F	31	037		3F	63	077	SUB	5F	95	137		7F	127	177	"

Table 3.2

EBCDIC Code Chart (2 of 2)

Control				Nymbers/Symbols				Upper case				Lower case			
Radix			EBCDIC	Radix			EBCDIC	Radix			EBCDIC	Radix			EBCDIC
Hex	Dec	Oct	Char	Hex	Dec	Oct	Char	Hex	Dec	Oct	Char	Hex	Dec	Oct	Char
80	128	200		A0	160	240		C0	192	300	{	E0	224	340	\
81	129	201	a	A1	161	241		C1	193	301	A	E1	225	341	
82	130	202	b	A2	162	242	s	C2	194	302	B	E2	226	342	S
83	131	203	c	A3	163	243	t	C3	195	303	C	E3	227	343	T
84	132	204	d	A4	164	244	u	C4	196	304	D	E4	228	344	U
85	133	205	e	A5	165	245	v	C5	197	305	E	E5	229	345	V
86	134	206	f	A6	166	246	w	C6	198	306	F	E6	230	346	W
87	135	207	g	A7	167	247	x	C7	199	307	G	E7	231	347	X
88	136	210	h	A8	168	250	y	C8	200	310	H	E8	232	350	Y
89	137	211	I	A9	169	251	z	C9	201	311	I	E9	233	351	Z
8A	138	212		AA	170	252		CA	202	312		EA	234	352	
8B	139	213		AB	171	253		CB	203	313		EB	235	353	
8C	140	214		AC	172	254		CC	204	314		EC	236	354	
8D	141	215		AD	173	255		CD	205	315		ED	237	355	
8E	142	216		AE	174	256		CE	206	316		EE	238	356	
8F	143	217		AF	175	257		CF	207	317		EF	239	357	
90	144	220		B0	176	260		D0	208	320	}	F0	240	360	0
91	145	221	j	B1	177	261		D1	209	321	J	F1	241	361	1
92	146	222	k	B2	178	262		D2	210	322	K	F2	242	362	2
93	147	223	l	B3	179	263		D3	211	323	L	F3	243	363	3
94	148	224	m	B4	180	264		D4	212	324	M	F4	244	364	4
95	149	225	n	B5	181	265		D5	213	325	N	F5	245	365	5
96	150	226	o	B6	182	266		D6	214	326	O	F6	246	366	6
97	151	227	p	B7	183	267		D7	215	327	P	F7	247	367	7
98	152	230	q	B8	184	270		D8	216	330	Q	F8	248	370	8
99	153	231	r	B9	185	271		D9	217	331	R	F9	249	371	9
9A	154	232		BA	186	272		DA	218	332		FA	250	372	
9B	155	233		BB	187	273		DB	219	333		FB	251	373	
9C	156	234		BC	188	274		DC	220	334		FC	252	374	
9D	157	235		BD	189	275		DD	221	335		FD	253	375	
9E	158	236		BE	190	276		DE	222	336		FE	254	376	
9F	159	237		BF	191	277		DF	223	337		FF	255	377	EO

Table 3.3

26

CHAPTER IV
HARDWARE ARCHITECTURE

- *BUS*

A *BUS,* or *Bus*, is a collection of parallel wires (connections or traces on *PCB*) through which electrical (*voltage*) or electronic signals (*data*) is transmitted from one part of a computer system to another. It is a single path or multiple parallel paths for power, control or data signals to which a device or several devices may be connected at the same time. In a computer system, the Bus is also known as the *Internal Bus* or *PC Bus*. It is a *Bus* that connects all internal computer devices and components to the CPU and memory. There is also the expansion Bus that enables the expansion boards to access the CPU and memory. Computers have three main BUS architectures: *Control BUS*, *Address BUS* and *Data BUS*.

The *Control BUS* consists of electrical signals (Clock, Read, Write, etc...) that permit the CPU to communicate with I/O devices and the memory. The *Address BUS* is an identifying number (often hexadecimal or octal) that describes a location in computer memory where Data BUS or information is stored. The *Address BUS* transfers the information to a specified location, i.e. to where the data should go. The *Data BUS*, which is bi-directional, transfers the actual data. *Control* and *Address BUSes* are usually uni-directionals.

The size of a *BUS*, known as the *width,* is important because it determines how much data can be transmitted at one time. For example, a 16-bit Bus can transmit 16 bits of data at once, whereas a 32-bit Bus can transmit 32 bits of data at once (twice as much).

Every Bus has a clock speed measured in *Megahertz* (*MHz*). A fast Bus allows data to be transferred faster, which makes the application software run faster. In Personal Computers, faster PCI Bus is replacing the original slow ISA Bus.

Most Personal Computers include a Local Bus for data that requires especially fast transfer speeds, such as video data. The local Bus is a high-speed pathway that connects directly to the *Central Processor Unit.*

Since the introduction of the digital circuit design, industry has tried to standardize on Bus structure but has failed to agree on a common Bus. The most used Bus structures are **PC Bus**, **Multibus**, **VAX**, **Q-Bus**, **VME Bus** and **STD Bus**.

- *Central Processor Unit (CPU)*

The *CPU*, also known as the *Processor* or the *Central Processor*, is the brain of the computer system (*Figure 4.0*: *(a)* Cyrix 6x86 P166+, Intel '486DX-33 and AMD '486DX-40; *(b)* pin-out of the Cyrix 6x86 P166+ CPU). The *CPU* must perform all of the computer's functions, decisions, executions and calculations. In terms of computing power, the *CPU* is the most important component of the computer system. The peripherals and devices that exist inside the computer are to bridge the gap between the *CPU* and the computer user. Before the Personal Computer era, the Mainframe computers, the *CPU* required one or more *Printed Circuit Board (PCB).*

(a)

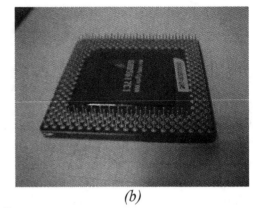

(b)

Figure 4.0

In the early 1980, the *Personal Computer* era began. There was a single IC called a *microprocessor* that was doing all the computer functions. The microprocessor is a single chip containing arithmetic and logic circuitry as well as control capability for memory and Input/Output access. The microprocessor controls the arithmetic and logic operations as well as the sequence of operations. It controls the storage of data, instructions, intermediate and final results of processing. In the late 1980's, the co-processor was introduced to assist and to relieve the *CPU* of part of its burden. In the 1990's, many co-processors (video, sound, keyboard, etc...) exist inside the *Personal Computer* helping the main *CPU* execute the application tasks. Most *CPU*s have two typical components, they are:

- The *Arithmetic Logic Unit (ALU)*, which performs arithmetic and logical operations.
- The *Control Unit* extracts program instructions from memory (ROM and/or RAM), decodes and executes them, calling on the *ALU* when necessary.

The 4004, *Intel's* first microprocessor, was introduced in 1971. The 4004 CPU was designed to process data arranged as 4-bit Data Bus and contains about 2,300 transistors. The *Intel* 4004 was intended to be used in a calculator. In 1974, *Intel* introduced the second generation of microprocessors: 8008, then the 8080 and the 8085. These CPUs were enhanced 4004 CPU, they were 8-bit Data Bus. In the late 1970's, most computer systems based on the then popular *CP/M Operating System* used the *Intel's* 8080, 8085 or the Zilog's Z-80.

The first *IBM Personal Computer* (PC) was based on *Intel's* 8088 *CPU*, a 16-bit microprocessor. This was followed by many *Intel* microprocessors such as 8086 (XT PC generations), 80286 ('286 PC generations), 80386 ('386 PC generations), 80486 ('486 PC generations), and the *Pentium* generations (*Pentium I, II* and *III*). In a span of about 19 years (1980 – 1999); the computer's performance with the 8080 *CPU* was 2 *MHz* (*Megahertz*), the 8088 *CPU* increased it to 4.77 *MHz*, the 80486 *CPU* made it as fast as 133 *MHz*, the *Pentium II* reached about 400 MHz and the *Pentium III* is expected to surpass to the 650 *MHz* (*Table 4.0*). For *AMD* and *Cyrix* microprocessors, refer to *Appendix A*.

The *Intel* processors (8088 through 80486) were designed primarily on *Complex Instruction Set Computing (CISC)* format. *CISC* uses command instructions that carry out single operation through many software instructions.

Intel Family of CPUs					
IC	Year	Data BUS Width (bits)	Address BUS width (bits)	Clock Speed (MHz)	No. of Transistors
8080	1974	8	8	2	6,000
8086	1978	16	20	5-10	29,000
8088	1979	8	20	4.77-8	29,000
80286	1982	16	24	6-25	134,000
80386SX	1988	32	24	16-33	275,000
80386DX	1985	32	32	16-40	275,000
80486SX	1991	32	32	16-33	1,185,000
80486DX	1989	32	32	25-66	1,200,000
80486DX2	1991	32	32	33-66	2,000,000
80486DX4	1992	32	32	75-100	2,500,000
Pentium	1993	32	32	60-166	3,300,000
Pentium Pro	1995	64	32	150-200	5,500,000
Pentium Pro II	1997	64	64	233-300	7,500,000

Table 4.0

Note: The *Advanced Micro Devices' (AMD) K6+ 3D CPU* has about 21.3 million transistors and there are approximately 36.5 million transistors in the *Intel Pentium Pro* with 512K *Cache*.

- *Low Insertion Force (LIF) socket*

The typical CPU socket is a *LIF* socket (*Figure 4.1*: *LIF* for a CPU and a co-processor). The CPU chip is inserted in this type of a socket with a little eccentric force until the chip is fully in place. It is a little more difficult to remove but not impossible. To remove or replace the CPU chip, one would need anti-static protection and a special removal tool or a small flat bladed screwdriver.

LIF

Figure 4.1

29

- *Zero Insertion Force (ZIF) socket*

The problem that would occur with the *LIF* (primarily bending then breaking a CPU pin(s) - e.g., CPU is damaged) resulted in the development of the *ZIF*. The *ZIF* is an IC socket that allows the insertion and removal of an IC chip without special tools (*Figure 4.2*: left – *ZIF* Socket 3 and right – *ZIF* socket 7).

Figure 4.2

- *Form Factor*

The *form factor* is the physical size and shape of a device. The major microprocessor chips manufacturers (*Intel*, *Advanced Micro Devices* and *Cyrix*) were using a common CPU pin-out. It is a pin-out configuration and circuit board size that was agreed by the major CPU chips manufacturers. This in turn will help computer developers to use the same motherboard for any of the CPUs. The advantage is for users to choose a CPU and upgrade that on the same motherboard. Different configurations were developed to meet the specifications for CPUs, socket were given a number as the new CPUs were introduced (*Table 4.1*). Refer to *Appendix B* for socket layout.

Socket Number	No. of Pins	Pin Layout	Supported Processors
Socket 1	169	17x17 PGA	SX/SX2, DX/DX2, DX4 OverDrive
Socket 2	238	19x19 PGA	SX/SX2, DX/DX2, DX4, OverDrive, 486 Pentium OverDrive
Socket 3	237	19x19 PGA	SX/SX2, DX/DX2, DX4, 486 Pentium OverDrive
Socket 4	273	21x21 PGA	Pentium 60/66, Pentium 60/66, OverDrive
Socket 5	320	37x37 SPGA	Pentium 75-133, Pentium 75+ OverDrive
Socket 6	235	19x19 PGA	DX4, 486 Pentium OverDrive
Socket 7	321	37x37 SPGA	Pentium 75-200, Pentium 75+ OverDrive
Socket 8	387	dual-pattern SPGA	Pentium Pro
SEC	242	Single Edge Cart. Slot	Pentium II

Note: Socket 6 was defined but never implemented in any systems.
SPGA: Staggered PGA

Table 4.1

- *Socket 3*

Socket 3 is the *form factor* for fifth-generation CPU chips such as '486 CPU generation from *Intel*, *AMD* and *Cyrix*. (*Figure 4.2*).

Figure 4.2

- *Socket 7*

Intel developed the *Socket 5* infrastructure. This was modified by other manufacturers, among them *AMD*. They renamed it *Socket 7*. It is the *form factor* for fifth-generation CPU chips such as *Intel's Pentium* series, *AMD K6* series and *Cyrix* CPU. It is a pin-out configuration and circuit board size that was agreed by the major CPU chips manufacturers (*Figure 4.3*). It is a 296-pin *ZIF* socket with connections for the CPU. The *Socket 7 Bus* operates at 66 MHz. The

Bus bandwidth is 533 *Mbps* at 66 *MHz* or 800 *Mbps* at 100 *MHz Bus*. The maximum *Data Bus* width is 64-bits.

Figure 4.3

In 1997, *Intel* introduced the first generation of CPUs with *MultiMedia Expansion* known as *MMX*. These CPUs included a set of 57 multimedia instructions. The *MMX*-enabled CPUs could handle many common multimedia operations, such as the *Digital Signal Processing (DSP),* that normally were controlled by a separate *Sound* or *Video* card. The 57 new instructions that have been added to the CPU are designed specifically to handle video, audio, and graphics data. However, special software (in *Basic Input/Output System (BIOS)* or application programs) to call the *MMX* instructions (inside the CPU) must be written to activate or take advantage of this feature. *MMX* incorporates a process that *Intel* named the **Single Instruction Multiple Data (SIMD)**. This process allows an instruction to perform the same function on many pieces of data which cuts down the repetitive loops of instructions that consume vast amounts of time to execute. The *SIMD* was used primarily as performance enhancements of *MMX* CPUs. The *SIMD* allows one *microinstruction* to operate at the same time on multiple data items. This is especially productive for applications where visual images or audio files are processed. A repeated succession of instructions (or loop) can now be performed in one instruction.

In 1998, *AMD* and other industry leaders introduced an enhanced *Socket 7* infrastructure and called it the *Super 7*. Among the enhancements for the *Super 7* are:

a) 100MHz local Bus frequency interface to the frontside Level 2 (L2) cache and main memory will speed up access to the cache and main memory by 50%. -Fast CPU response.

b) *Accelerated Graphics Port (AGP)* support. -Fast graphics.

c) Backside L2 cache and Frontside L3 cache support. -High performance.

- *Socket 8*

Socket 8 is the *form factor* for *Intel's Pentium Pro* CPU. It is a 387-pin *ZIF* socket with connections for the CPU and one or two *SRAM* die for the level 2 Cache.

- *Slot 1*

Slot 1, Intel's proprietary, is the *form factor* for *Intel's Pentium II* (code named *Klamath*) and *Pentium III* CPU. The Slot 1 replaces the *Socket 7* and *Socket 8* form factors used by *Intel's* previous *Pentium* Processors, i.e. CPU on the motherboard is not backward compatible. Slot 1 is a 242-pin edge connector contact daughter-card slot that accepts a microprocessor packaged as a *Single Edge Contact*, or *SEC* cartridge. A typical motherboard can have one or two Slot 1s (*Figure 4.4*). The Slot 1 Bus operates at 66 MHz. The Bus bandwidth is 533 Mbps at 66 MHz or 800 Mbps at 100 MHz Bus. The maximum Data Bus width is 64-bits.

With the Introduction of the *Slot 1* infrastructure, Intel introduced the **Dual Independent Bus** (**DIB**) architecture. This Bus supports two independent Buses unlike the previous infrastructures (*Socket 1* through *Socket 7*) that had a single Bus architecture; *Socket 8* supports *DIB*. The reason for the two independent Buses is one for accessing the *L2* cache and one for main memory. These Buses can operate simultaneously which accelerates the flow of data within the system.

Figure 4.4

- *Clock speed*

The **clock speed**, also referred to as **clock rate**, is the speed at which a CPU of a computer executes program instructions. Every computer system contains an internal clock that regulates the rate at which instructions are executed and synchronizes all the various computer components. The CPU requires a fixed number of clock ticks (or **clock cycles**) to execute each instruction. The faster the clock, the more instructions the CPU can execute per second. Clock speeds are measured in *MegaHertz* (*MHz*), 1 MHz being equal to 1 million cycles per second. The CPUs of Personal Computer systems have clock speeds of anywhere from 2 *MHz* to over 400 *MHz* (*Table 4.0*). The internal architecture of a CPU has as much to do with a CPU's performance as the clock speed; so two CPUs with the same clock speed will not necessarily perform equally. Whereas an *Intel* 80286 microprocessor requires 20 cycles to multiply two numbers, an Intel 80486 or later Processor can perform the same calculation in a single clock tick. These newer processors, therefore, would be 20 times faster than the older processors even if their clock speeds were the same. The CPU continuously executes the cycle events shown in *Figure 4.5*. This cycle is often called the *"instruction cycle"* of the CPU.

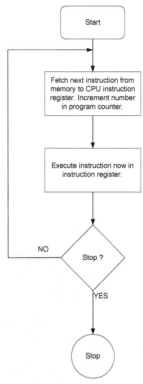

Figure 4.5

In addition, some microprocessors are *Superscalar*, which means that they can execute more than one instruction per clock cycle. Like CPUs, expansion *Buses* also have clock speeds. Ideally, the CPU clock speed and the *Bus* clock speed should be the same so that neither component slows down the other. In practice, the *Bus* clock speed is often slower than the CPU clock speed, which creates a bottleneck. This is why new local buses, such as *PCI* Bus and *AGP* Bus, have been developed.

- *Million Instructions Per Second (MIPS)*

The *MIPS* is a measurement used for the computer's speed and power. *MIPS* measures roughly the number of machine instructions that a computer can execute in one second. Different instructions, however, require more or less time than others, and there is no standard method for measuring *MIPS*. In addition, *MIPS* refers only to the CPU speed, whereas real software applications are generally limited by other factors, such as I/O speed (drive access) and others. A machine with a high *MIPS* rating, therefore, might not run a particular application any faster than a machine with a low *MIPS* rating. For all these reasons, *MIPS* ratings are not used often anymore. Despite these problems, a *MIPS* rating can give the user a general idea of a computer's speed. The *IBM PC/XT* computer, for example, is rated at ¼ *MIPS*, while the *Pentium*-based PC runs at over 100 *MIPS*.

- *Benchmark*

A *Benchmark* is a test used to compare performance of the hardware and/or software. Many trade magazines have developed their own *benchmark* tests, which they use when reviewing a class of products. When comparing *benchmark* results, it is important to know exactly what the

34

benchmarks are designed to test. A *benchmark* that tests graphics speed focuses primarily on video and graphics tests.

- *Access Time*

The *Access Time* is the time that it takes a software program or a device to locate information and make that information available to the computer's CPU for processing (*Table 4.2*). For example, the time measured for the CPU to retrieve data from memory is the *access time*. Ideally, the *access time* of memory should be as fast or faster than the CPU to keep up with the requests. If not, the CPU will waste a number of clock cycles that will cause the computer to be slower. The reported *access time*, however, is sometimes misleading because most memory chips especially the DRAM, require a pause between back-to-back accesses. This is one reason that the SRAM is much faster than DRAM. SRAM does not require any Refresh cycle, hence, there is no pause between back-to-back accesses.

Typical Access Time (seconds)	Device
5 to 15 nano	Static RAM or SRAM.
50 to 150 nano	Dynamic RAM or DRAM.
55 to 250 nano	EPROM.
55 to 250 nano	ROM.
9 to 15 milli	Hard Disk Drive or HDD.
19 to 100 milli	Erasable Optical Drive.
80 to 800 milli	CD ROM Drive.
approx. 20	DAT Tape Drive.
approx. 40	QIC Tape Drive.
40 to 500	8 mm Tape Drive.

Table 4.2

The *access time* of a disk drive includes the time it takes for the Read/Write Head to locate a sector on the disk (also called *Seek Time*). This time is averaged because it depends on the distance the head has to travel to locate the desired data. Note that the access time of a hard disk drive is at least 200 times slower than the average DRAM access time.

- *Co-Processor*

The *co-processor* is a special-purpose-processing unit that assists the CPU in performing certain types of operations. For example, a *math co-processor* performs mathematical computations, particularly *Floating-Point Operations*. Math coprocessors are also called *numeric* and *Floating-Point co-processors*. Most computer systems come with *Floating-Point co-processors* built in; the firmware and software program must be written to take advantage of the performance. If the program does not contain co-processor instructions, the co-processor will not be utilized.

In addition to math co-processors, there are also *Graphics* co-processors for manipulating graphic images. These are often called *Accelerator* boards.

- *Floating Point Number (FPN)*

The *FPN* is a real number (that is, a number that can contain a fractional part). The following are floating-point numbers:

35

1.0

-51.3

6½

4E-2

The term *Floating-Point* is derived from the fact that there is no fixed number of digits before and after the decimal point; that is, the decimal point can float. There are also representations in which the number of digits before and after the decimal point is set, called Fixed-Point representations. In general, Floating-Point representations are slower and less accurate than Fixed-Point representations, but they can handle a larger range of numbers.

Note that most floating-point numbers a computer can represent are just approximations. One of the challenges in programming with floating-point values is ensuring that the approximations lead to reasonable results. If the programmer is not careful, small discrepancies in the approximations can snowball to the point where the final results become meaningless. Because mathematics with floating-point numbers requires a great deal of computing power, many microprocessors come with an FPU chip.

o *Floating Point Unit (FPU)*

The *FPU* is a specially designed chip that performs Floating-Point calculations. Computer systems that are equipped with an *FPU* perform certain types of applications much faster than computers that lack one. In particular, graphics applications are faster with an *FPU*. Some microprocessors, such as the Intel 80486 and Pentium, have a built-in *FPU*. With other microprocessors, an *FPU* can be added by inserting the *FPU* chip on the motherboard. Floating-Point Units are also called numeric co-processors, math coprocessors and floating-point processors.

o *FLoating-point OPerations per Second (FLOPS)*

The *FLOPS* is a common benchmark measurement for rating the speed of computer microprocessors. Floating-point operations include any operations that involve fractional numbers. Such operations, which take much longer to compute than integer operations, occur often in some applications.

Most new microprocessors include a *FPU*. The *FLOPS* measurement, therefore, actually measures the speed of the *FPU*. The *MegaFLOPS (MFLOPS)* is equal to one million floating-point operations per second, and the *GigaFLOPS (GFLOPS)* is equal to one billion floating-point operations per second.

CHAPTER V
MAIN HARDWARE

- *Interrupt Request (IRQ)*

The *IRQ* is a system resource. Every computer has interrupts (*Table 5.0*) to get the attention of the *CPU*, as needed, example: there are interrupts connected to the keyboard, the mouse, the hard drive, the floppy drive, etc... As the *CPU* is executing a certain operation of a function, to keep track of what it is doing before a device interrupts it; the *CPU* writes the address of the operation or a function into a special memory location in *Random Access Memory (RAM)* called the *Stack*.

Interrupt Controller			
Interrupt	*Device Assigned*	*Typical usage*	*Device Available ?*
0	System Timer Interrupt.	System Clock	NO
1	Keyboard Controller.	Keyboard	NO
2	Cascade for IRQ 8 to IRQ 15.	Cascade for IRQ 8 to IRQ 15.	Yes: Network card
3	Serial Port 2.	Serial Port: COM 2 and COM 4	Yes: Modem or mouse
4	Serial Port 1.	Serial Port: COM 1 and COM 3	Yes: Modem or mouse
5	Parallel Port 2.	Parallel Port: LPT2	Yes: Sound or Network card
6	Floppy Disk Controller.	Floppy Disk Controller.	NO
7	Parallel Port 1.	Parallel Port: LPT1	Yes: LPT1 printer
8	Clock / Calendar.	Real time clock	NO
9	Available.	Linked to IRQ 2	Yes: Network card
10	Available.	Available.	Yes: CD ROM or Network card
11	Available.	Available.	Available
12	Available.	PS/2 mouse	Available
13	Math Co-Processor.	Math Co-Processor.	Available
14	Available.	Primary Hard Disk Controller	NO
15	Available.	Secondary Hard Disk Controller	Available

Table 5.0

When a device needs the *CPU*'s attention, the interrupt controller notifies it through an electrical signal. The execution of the operation is temporary stopped and the *CPU* attention is diverted to service this interrupt signal. The recognition of the interrupt source and the priority can be retrieved from the interrupt controller. The *CPU* checks the interrupt table in *RAM* memory (*Stack*), reads and executes the instructions in *Basic Input/Output System (BIOS) ROM* memory, then generates an *Interrupt RETurn (IRET)* instruction. The *CPU* now can retrieve the address from the Stack and return to its initial operation or function.

o *Interrupt Priority*

The PC processes device interrupts according to their priority level. This is a function of which interrupt line they use to enter the interrupt controller. For this reason, the priority levels are directly tied to the interrupt number:

- On a *PC/XT*, the priority of the interrupts is 0, 1, 2, 3, 4, 5, 6 and 7.
- On a new PC, it is slightly more complicated. The second set of eight interrupts is piped through the *IRQ2* channel on the first interrupt controller. This means that the first controller views any of these interrupts as being at the priority level of its "*IRQ2*". The result of this is that the priorities become 0, 1, (8, 9, 10, 11, 12, 13, 14, 15), 3, 4, 5, 6 and 7. i.e., *IRQs* 8 to 15 take the place of *IRQ2*.

The priority level of the *IRQs* does not make much of a difference in the performance of the machine. Higher-priority IRQs may improve the performance of the devices that use them slightly.

- *Direct Memory Access (DMA)*

The *DMA* is a system resource that assists computer components (video, sound cards, etc.) in communication with the *CPU* and other peripherals within the computer system. The *DMA* (*Table 5.1*) is a direct line from the CPU / the peripherals to the memory. It is the process of one intelligent device accessing the memory of another, bypassing the other's processor's internal memory. The maximum rate for *DMA* is 16.6 *Mbps*.

DMA channel	Bus Line	Typical Default Use	Other Common Uses
0	no	Memory Refresh	None
1	8/16-bit	Sound card (low DMA)	SCSI host adapters, ECP parallel ports, tape accelerator cards, network cards, voice modems
2	8/16-bit	Floppy disk controller	Tape accelerator cards
3	8/16-bit	None	ECP parallel ports, SCSI host adapters, tape accelerator cards, sound card (low DMA), network cards, voice modems, hard disk controller on old PC/XT
4	no	None; cascade for DMAs 0-3	None
5	16-bit only	Sound card (high DMA)	SCSI host adapters, network cards
6	16-bit only	None	Sound cards (high DMA), network cards
7	16-bit only	None	Sound cards (high DMA), network cards

Table 5.1

Quantum Corporation and *Intel Corporation* developed the *Ultra DMA* as a proposed industry standard. The *Ultra DMA (or Ultra DMA/33)* is a protocol for transferring data between a hard disk drive through the computer's *Data Bus*, to the computer's *RAM* memory. This protocol transfers data in burst mode at a rate of 33.3 Mbps, twice as fast as the previous *DMA* interface. An *Ultra DMA* support in a computer means that it will start and open new applications more quickly. It will also help users of graphics-intensive and other applications that require large amounts of access to data on the hard drive. *Ultra DMA* uses *Cyclical Redundancy Checking (CRC),* that checks for data transfer errors, offering a new level of data protection. Because the *Ultra DMA* protocol is designed to work with legacy *Programmed Input/Output (PIO)* and *DMA* protocols, it can be added to many existing computers by installing an *Ultra DMA/33 PCI* adapter card. *Ultra DMA* uses the same 40-pin *IDE* interface cable as existing *PIO* and *DMA.*

- *Programmed Input/Output (PIO)*

The *PIO* is a method of transferring data between two devices that uses the computer's main processor as part of the data path. *PIO* is a way of moving data between devices in a computer in which all data must pass through the processor. A newer alternative to *PIO* is the *DMA*. As expected, *PIO* was dropped from industry standards and was replaced entirely by *DMA* and *Ultra DMA. ATA/IDE* standard uses *PIO* and defines the speed of the data transfer in terms of the *PIO* mode implemented, as shown in *Table 5.2.*

PIO Mode	*Data Transfer Rate (MBps)*	*Standard*
0	3.3	ATA
1	5.2	ATA
2	8.3	ATA
3	11.1	ATA-2
4	16.6	ATA-2

Table 5.2

- *AT Attachment (ATA)*

The *ATA* is a disk drive implementation that integrates the controller on the disk drive itself. There are several versions of *ATA*, all developed by the *Small Form Factor (SFF) Committee:*

  o *ATA:* or *Integrated Device Electronics (IDE),* supports one or two hard drives, a 16-bit interface and *PIO* modes 0, 1 and 2.
  o *ATA-2:* Supports faster *PIO* modes (3 and 4) and multiword *DMA* modes (1 and 2). Also supports *Logical Block Addressing (LBA)* and block transfers. *ATA-2* is marketed as *Fast ATA* and *Enhanced IDE (EIDE).*
  o *ATA-3:* Minor revision to ATA-2.
  o *Ultra-ATA:* Also called *Ultra-DMA, ATA-33,* and *DMA-33,* supports multiword *DMA* mode 3 running at 33 Mbps.
  o *ATA/66:* A new version of *ATA* proposed by *Quantum Corporation,* and supported by *Intel*, that will double *ATA's* throughput to 66 Mbps.

- *AT Attachment Packet Interface (ATAPI)*

The *ATAPI* is an extension to *EIDE* (also called *ATA-2*) that enables the interface to support *CD-ROM* players and tape drives.

- *Motherboard (Main board)*

The *motherboard* (*Figure 5.2*) is the main circuit board of the computer. The motherboard contains the connectors for attaching additional expansion boards. Typically, the motherboard contains the *CPU*, *BIOS*, Memory (*RAM*), mass storage interfaces, serial ports (*RS-232C*), parallel port (*LPT1*), expansion slots, and all the controllers required to control standard peripheral devices, such as the display screen, keyboard and disk drive. Collectively, all these chips that reside on the motherboard are known as the motherboard's chipset. On most computer systems, there are open slots to add memory chips directly to the motherboard.

Figure 5.2

Motherboards are designed in different formats:

1. *Backplane systems* use a main card with slots to plug all of the PC components into. These systems typically reside in industrial application due to the cost. There are two types of *backplane systems*:

   a. *Passive*: The *Passive backplane system* has a main card with slots mounted on the card and no circuitry on the card. Each piece of the PC is mounted on its own card and then plugged into the backplane.

   b. *Active*: The *Active backplane system* has some circuitry mounted on the backplane but not a CPU. The card that has the CPU mounted on it is called the processor complex.

   This setup allows the user to swap or upgrade components very quickly. Unfortunately, there is no standard for the cards that plug into the backplane so the computer system is limited to what the specific manufacturer sells. The *backplane systems* are all rack mounted (CPU board and all peripherals are plug-ins cards on a common *PCB*) and the system Bus is fixed.

2. The *Full-Size AT* is a form that matches the original *IBM AT* motherboard design. The board dimensions are 12 inches wide by 13.8 inches deep. Due to advances in

computer miniaturization and the fact that this form will not fit into the popular Baby-AT chassis, this form is no longer produced.

3. The *Baby AT* is the form factor used by most PC motherboards prior to 1998. The original motherboard for the *PC/AT* measured 12 inches wide by 13.8 inches deep. Baby AT motherboards are a little smaller, 8.5 inches wide by 13 inches deep usually. The *Baby AT* form is popular because it fits into virtually all *"full size"* AT cases and of course the *Baby-AT* case. This form has specific placement of the keyboard connector and also the expansion slot connectors. The only case that this form does not fit into is the Low profile of *"Slimline"* cases.

4. The *ATX* (*Figure 5.3*) is the new shape and layout of PC motherboards. It improves on the previous standard, the *Baby AT form factor,* by rotating the orientation of the board 90 degrees. This allows for a more efficient design, with disk drive cable connectors nearer to the drive bays and the *CPU* closer to the power supply and cooling fan.

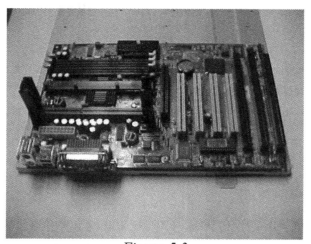

Figure 5.3

The *ATX* form is an advanced design over the previous *AT* style motherboards. The form is not compatible with any other form. A new chassis was designed to accommodate the new form. The official specification was released by *Intel* in 1995 and has recently been revised to version 2.01 in February of 1997. Several major areas have been improved over previous forms such as: a stacked I/O connector panel mounted on the motherboard, cables internally are not needed for the serial COM ports and a single keyed internal power supply connector. This makes it easier for the user to plug the power supply into the motherboard. Another improvement is the relocation of the CPU and memory. They do not interfere with any *Bus* expansion slots. Improved cooling is another added benefit. The improvements lowered the cost to manufacturer.

5. The *LPX* is a motherboard form factor that is used in some desktop PCs. The distinguishing characteristic of *LPX* is that expansion boards are inserted into a *riser* that contains several slots. So the expansion boards are parallel to the motherboard rather than perpendicular to it as in other common form factors, such as *AT* and *ATX*.

The *LPX* design allows for smaller cases, but the number of expansion boards is usually limited to two or three. Western Digital developed the LPX Form for some of their motherboards. *Western Digital* no longer produces this motherboard but other manufacturers have adopted it. The *LPX* and *Mini-LPX* are small and typically cheaper to make. The main distinctive feature is the riser card that contains all of the expansions slots. Another feature is the placement of the video circuit, parallel port, two serial ports, and mini-DIN PS/2 style Mouse and Keyboard connectors.

6. The *NLX* is also a new form factor designed by *Intel Corporation* for PC motherboards. The *NLX* form factor features a number of improvements over the current *LPX* form factor and is expected to be widely implemented starting in 1998. Its features include:

   - Support for larger memory modules and *Dual In-line Memory Module (DIMM)*.
   - Support for the newest microprocessors, including the *Pentium II* using *Single Edge Contact (SEC)* packaging.
   - Support for *Accelerated Graphics Port (AGP)* video cards.
   - Better access to motherboard components
   - Support for *dockable* designs in which the motherboard can be removed without tools.

NLX is the latest technology in motherboard *Form Factor*. The concept is similar to *LPX* but with a number of improvements. Some of the improvements are:

   o Support for current processor technologies such as the *Pentium II* and beyond.

   o The flexibility to rapidly change processor technologies without tearing the computer system apart.

The new *Form* also has support for emerging technologies which include *Accelerated Graphics Port (AGP), Universal Serial Bus (USB),* and tall memory modules.

CHAPTER VI
EXPANSION ARCHITECTURE

- *Expansion slot*

The *expansion slot* is an opening in a Personal Computer where a circuit board can be inserted to add new capabilities to the computer. Most Personal Computers except portables and laptops contain expansion slots for adding more graphics, I/O capabilities, and support for special devices. The Printed Circuit Boards inserted into the expansion slots are called expansion boards. Expansion slots for Personal Computers come in two basic sizes: *full-size (Figure 6.0)* and *half-size (Figure 6.1)*.

Figure 6.0

- *Expansion board*

The *expansion board* (also called card, add-in and add-on) is a printed circuit board that can insert into a motherboard slot to give the computer system added capabilities. Examples, of expansion boards are:

- Video adapter card.
- Video Graphics Accelerator card.
- Sound card.
- CPU accelerator board.
- Internal modem card.

The expansion board of a Personal Computer system can be *half-size* (*half-length*) or *full-size* (*full-length*). Most Personal Computers have slots for each type of board. A half-size *Industry Standard Architecture (ISA)* board is sometimes called an *8-bit board* (*Figure 6.1*) because it can transmit only 8 bits at a time, but a *Peripheral Component Interconnect (PCI)* board is either 16-bit or 32-bit board. A full-size *ISA* board is called a *16-bit board* (*Figure 6.2*).

Figure 6.1

Figure 6.2

- *Daughter board*

A *daughter board* (also called *daughter card*) is a Printed Circuit Board that plugs into another circuit board (usually the main board or motherboard). A daughter board is similar to an expansion board, but it accesses the motherboard components (CPU and Memory) directly instead of sending data through the slower expansion Bus.

- *Adapter*

An *adapter* is a physical device that allows one hardware or electronic interface to be adapted (accommodated without loss of function) to another hardware or electronic interface. In a computer, an *adapter* is often built into a card that can be inserted into a slot on the computer's motherboard. The card adapts information that is exchanged between the computer's microprocessor and the devices that the card supports.

- *Accelerator board*

There are two types of *Accelerator board:*

1. Graphics accelerator board that rapidly refreshes and updates the video/graphics on the monitor of the PC.
2. CPU accelerator board that makes the PC faster.

Most modern computers have fast CPUs (400 MHz or faster) and *Accelerated Graphics Port (AGP)* video cards. Most accelerator boards were developed and used between 1985 and 1995.

- *Peripheral Device*

The peripheral device is an input/output device that is attached to a computer system internally or externally. PC cards (*ISA, PCI, SCSI*), storage devices (floppy drive, hard drive, CD ROM drive) and CRT monitor are examples of peripheral devices.

- *Industry Standard Architecture (ISA)*

The *ISA* Bus (*Figure 6.1*) was introduced in the early 1980's with the *IBM PC, XT* and *PC/AT* computer systems. The *AT* version of the bus is called the *AT Bus* and became a *de facto* PC Bus industry standard. Until early 1990, most *PC* cards were using this architecture. It was then replaced by the *PCI Local Bus* architecture. Most computer systems made today include both an AT bus for slower devices and a *PCI* bus for devices that need better bus performance. The data burst transfer of the original architecture is 2 *Megabits per second* (*Mbps*), later one were 8.33 *Mbps*. In 1993, the *Plug and Play ISA* was introduced enabling the *Operating System* to configure expansion boards automatically (PC users do not need to manually configure the PC boards with *DIP* switches and/or jumpers).

- *Video Electronic Standard Association (VESA)*

The *VESA* or *VESA local Bus* or *VESA VL Bus* or *VLB,* was introduced in 1992 (*Figure 6.3*). The first local bus to gain popularity, it was developed to allow high-speed connections between peripherals. *VESA Local Bus* is a standard interface between the computer and its expansion slots. It provides faster data flow between the devices controlled by the expansion cards and the computer's microprocessor. A *"Local Bus"* is a physical path on which data flows at almost the speed of the microprocessor, increasing total system performance. The *VLB* is a 32-bit Bus which is a direct extension of the '486 Processor/Memory Bus. A *VLB* slot is a *16-bit ISA* slot with third and fourth slot connectors added on the end. *VESA Local Bus* is particularly effective in systems with advanced video cards and supports 32-bit data flow at 50 MHz. A *VESA Local Bus* is implemented by adding a supplemental slot and card that aligns with and augments an *ISA* expansion card. The *VLB* normally runs at 33 MHz, although higher speeds are possible on some systems. Use of a *VLB* video card and I/O controller greatly increases system performance over an *ISA*-only system. While *VLB* was extremely popular during the reign of the '486 CPU, with the introduction of the Pentium and its *PCI Local Bus* in 1994, wholesale abandonment of the *VLB* began in earnest.

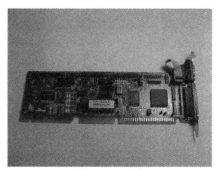

Figure 6.3

- *Extended Industry Standard Architecture (EISA)*

In the mid 1980's, the *EISA Bus* was introduced to meet the next generation of faster computer technology. This *Bus* architecture was designed for computer systems using *Intel* 80386, 80486, or *Pentium* CPUs. *EISA Bus* is 32-bit wide and support multiprocessing. The *EISA* Bus was designed by nine *IBM's* competitors (sometimes called the *Gang of Nine*): *AST*

Research, Compaq Computer, Epson, Hewlett-Packard, NEC, Olivetti, Tandy, WYSE, and *Zenith Data Systems*. They designed the architecture to compete with *IBM's* own high-speed bus architecture called the *Micro Channel Architecture* (*MCA*). The principal difference between *EISA* and *MCA* is that *EISA* is backward compatible with the *ISA* Bus (also called the *AT Bus*), while *MCA* is not. Computers with an *MCA* Bus can use only *MCA* expansion cards. *EISA* and *MCA* are not compatible with each other. Some of the key features of the *EISA Bus*:

- *ISA Compatibility: ISA* cards will work in *EISA* slots.

- *32 Bit Bus Width:* Like *MCA*, the *Bus* was expanded to 32 bits.

- *Bus Mastering:* The *EISA Bus* supports *Bus* Mastering adapters for greater efficiency, including proper Bus arbitration.

- *Plug and Play: EISA* automatically configures adapter cards, similar to the *Plug and Play* standards of modern systems.

- *Peripheral Component Interconnect (PCI)*

The *PCI* Bus (*Figure 6.4*), also known as *Local Bus Standard*, was developed by Intel and introduced in 1993. It was geared specifically to *Fifth-* and *Sixth-Generation Systems*, although the latest generation '486 motherboards use *PCI* as well. Like the *VESA Local Bus*, *PCI* is a 32-bit bus that normally runs at a maximum of 33 *MHz*. The key to *PCI*'s advantages over its predecessor, the *VESA local bus*, lies in the chipset that controls it. The *PCI Bus* is controlled by special circuitry in the chipset that is designed to handle it, where the *VLB* was basically just an extension of the '486 Processor *Bus*. *PCI* is not married to the '486 in this manner, and its chipset provides proper Bus arbitration and control facilities, to enable *PCI* to do much more than *VLB* ever could. *PCI* is also used outside the PC platform, providing a degree of universality and allowing manufacturers to save on design costs. *PCI* became the new computer industry standard *Bus*. *PCI* is an interconnection system between a microprocessor and attached devices in which expansion slots are spaced closely for high-speed operation. Using *PCI*, a computer can support both new *PCI* cards while continuing to support *ISA* expansion cards, currently the most common kind of expansion card. Designed by *Intel*, the original *PCI* was similar to the *Vesa Local Bus*. However, *PCI* 2.0 is no longer a local bus and is designed to be independent of microprocessor design. *PCI* is designed to be synchronized with the clock speed of the microprocessor, in the range of 20 to 33 *MHz*. *PCI* is now installed on most new desktop computers, not only those based on Intel's Pentium processor but also those based on the *PowerPC*. *PCI* transmits 32 bits at a time in a 124-pin connection (the extra pins are for power supply and grounding) and 64 bits in a 188-pin connection in an expanded implementation. *PCI* uses all active paths to transmit both *Address* and *Data* signals, sending the *Address* on one clock cycle and *Data* on the next. Burst data can be sent starting with an *Address* on the first cycle and a sequence of *Data* transmissions on a certain number of successive cycles. The *PCI Bus* architecture meets most *Fifth Generation Systems* computer technology primarily the *Plug & Play*. The data burst transfer of original *PCI* architecture is 133 *Mbps*; the new one will be 266 *Mbps*. Most modern personal computer systems include a *PCI* bus in addition to a more general *ISA* expansion *Bus*. *PCI* is a 64-bit *Bus*, but usually implemented as a 32-bit *Bus*. It can run at clock speeds of 33 or 66 *MHz*. At 32-bits and 33 *MHz*, it yields a throughput rate of 133 *Mbps*.

Figure 6.4

- *Peripheral Component Interconnect-X (PCI-X)*

A coalition of vendors led by *Compaq Computer Corporation, IBM* and *Hewlett-Packard* developed the *PCI-X Bus* to take advantage of Intel's faster microprocessors. The *PCI-X* is a *PCI* based I/O Bus but improved in performance with increased flow data between CPU and various peripherals. The *PCI-X* is a 64-bit Bus that runs at speeds up to 133 *MHz*. Theoretically, it will be capable of transmitting data at speeds exceeding the 1 *Gigabit per second (Gbps)*.

- *Accelerated Graphics Port (AGP)*

The need for increased bandwidth between the main processor and the video subsystem originally led to the development of the local *I/O Bus* on the PCs, starting with the *VESA local Bus* and eventually leading to the popular *PCI Bus*. This trend continues, with the need for video bandwidth now starting to push up against the limits of even the *PCI Bus*. Much as was the case with the *ISA Bus* before it, traffic on the *PCI Bus* is starting to become heavy on high-end PCs, with video, hard disk and peripheral data all competing for the same I/O bandwidth. To combat the eventual saturation of the *PCI Bus* with video information, *Intel*, having designed specifically for the video subsystem, has pioneered a new interface. It is called the *Accelerated Graphics Port* or *AGP*. *AGP* (*Figure 6.5*) is a new technology that was introduced in 1997. *Intel's* goal with *AGP* was to be a more affordable high-end video without requiring sophisticated 3D video cards. *AGP* is based on *PCI* Bus, but is designed especially for the throughput demands of 3-D graphics. Rather than using the *PCI* bus for graphics data, AGP introduces a dedicated point-to-point channel so that the graphics controller can directly access main memory. *AGP* was developed in response to the trend towards greater and greater performance requirements for video. As software evolves and computer use continues into previously unexplored areas such as 3D acceleration graphics and full-motion video playback, both the processor and the video chipset need to process more and more information. The *PCI* bus is reaching its performance limits in these applications, especially with hard disks and other peripherals also in there fighting for the same bandwidth. The *AGP* channel is 32-bit wide (the same as *PCI* Bus) and runs at 66 *MHz* (on a standard *Pentium II* motherboard instead of the *PCI Bus'* 33 *MHz*), it contains its own CPU to boost performance levels by computing in real time the graphical transformations from CPU onto the PC monitor. This translates into a total bandwidth of 266 *Mbps*; as opposed to the *PCI* bandwidth of 133 *Mbps*. *AGP* also supports two optional faster modes, with throughputs of 533 *Mbps* and 1.07 *Gbps*. In addition, *AGP* allows 3-D textures to be stored in main memory rather than video memory. The idea behind *AGP* is simple: create a faster, dedicated interface between the video chipset and the system processor. The interface is only between these two devices; this has three major advantages: it makes it easier to implement the port, it is easier to

Carlos E. Hattab

increase *AGP* in speed, and makes it possible to put enhancements into the design that are specific to video. *AGP* is considered a port, and not a *Bus*, because it only involves two devices (the processor and video card) and is not expandable. One of the great advantages of *AGP* is that it isolates the video subsystem from the rest of the PC so there is not nearly as much data contention over I/O bandwidth as there is with *PCI*. With the video card removed from the *PCI* bus, other *PCI* devices will also benefit from improved bandwidth.

AGP has some important system requirements:
- The chipset must support *AGP*.
- The mainboard must be equipped with an *AGP* bus slot or must have an integrated *AGP* graphics system.
- The Operating System must be the *Microsoft Windows 95 version B, Microsoft Windows 98* or *Windows 2000*.

Figure 6.5

With the popularity of the multimedia (graphics and audio) and graphical applications, the *AGP* can free up the computer's CPU to execute other commands while the graphics accelerator is handling graphics computations. Aside from the graphics processor used, the other characteristics that differentiate graphics accelerator board from the traditional video board is:
- Memory: Graphics accelerator board has its own memory, which is reserved for storing graphical representations. The amount of memory determines how much resolution and how many colors can be displayed. Some graphics accelerator boards use conventional *DRAM*, but others use a special type of *Video RAM (VRAM)*, which enables both the video circuitry and the processor to simultaneously access the memory.
- Bus: Graphics accelerator board is normally designed for a particular type of video bus. For example, as of 1995, most boards were designed for the *PCI* bus.
- Register width: The traditional *ISA* video board were either 8-bit or 16-bit data wide, the *PCI* were 32-bit wide. The wider the register, the more data the processor can manipulate with each instruction. Modern computers have 64-bit graphics accelerator boards and the 128-bit graphics accelerator boards are in the near future.
- *Plug and Play (PnP)*

The large variety of different cards that can be added to a PC in order to expand its capabilities is both a blessing and a curse. Configuring the system and dealing with resource conflicts (*IRQ, DMA, I/O* ports) is part of the curse of having so many different non-standard devices available on the market. Dealing with these issues can be a tremendously confusing, difficult and time-consuming task. In fact, many users have stated that this is the single most frustrating part of owning and maintaining a PC, or of upgrading the PC's hardware. In an attempt to resolve this ongoing problem, the *Plug and Play (PnP)* specification was developed in 1995 with the release of *Windows 95* and PC hardware designed to work with it, by *Microsoft*, with cooperation from *Intel* and many other hardware manufacturers. The goal of *PnP* is to create a computer whose hardware and software work together to automatically configure devices and assign resources, to allow for hardware changes and additions without the need for large-scale resource assignment tweaking. As the name suggests, the goal is to be able to just plug in a new device and immediately be able to use it, without complicated setup maneuvers. A form of *Plug and Play* was actually first made available on the *EISA* and *MCA* Buses many years ago. For several reasons, however, neither of these Buses caught on and became popular. Most of the actual work involved in making *PnP* function is performed by the system *Basic Input/Output System (BIOS)* during the boot process. At the appropriate step of the boot process, the *BIOS* will follow a special procedure to determine and configure the *Plug and Play* devices in the system. A rough layout of the steps that the *BIOS* follows at boot time when managing a *PCI*-based *Plug and Play* system:

1. Create a resource table of the available *IRQ*, *DMA* channels and I/O addresses, excluding any that are reserved for system devices.
2. Search for and identify PnP and non-PnP devices on the *PCI* and *ISA* buses.
3. Load the last known system configuration from the *Extended System Configuration Data (ESCD)* area stored in non-volatile memory.
4. Compare the current configuration to the last known configuration. If they are unchanged, continue with the boot; this part of the boot process ends and the rest of the bootup continues from here.
5. If the configuration is new, begin system reconfiguration. Start with the resource table by eliminating any resources being used by non-PnP devices.
6. Check the *BIOS* settings to see if any additional system resources have been reserved for use by non-PnP devices and eliminate any of these from the resource table.
7. Assign resources to *PnP* cards from the resources remaining in the resource table, and inform the devices of their new assignments.
8. Update the *ESCD* area by saving to it the new system configuration. Most *BIOS* will print a message when this happens like "Updating *ESCD* ... Successful".
9. Continue with the "boot".

• *Micro Channel Architecture (MCA)*

Introduced in 1987, *IBM* Developed the *MCA* or *Micro Channel Bus* (*Figure 6.6*), for its line of *PS/2* desktop computers. *MCA Bus* is an interface between a computer (or multiple computers) and its expansion cards and their associated devices. *MCA* was a distinct break from previous *Bus* architectures such as *ISA*. The pin connections in *MCA* are smaller than other *Bus* interfaces. For this and other reasons, *MCA* does not support other *Bus* architectures. Although *MCA* offers a number of improvements over other bus architectures, its proprietary, nonstandard aspects did not encourage other manufacturers to adopt it. It has influenced other *Bus* designs and it is still in use in *PS/2s* and in some minicomputer systems. The *MCA Bus* was *IBM's* attempt to replace the *ISA Bus* with something "bigger and better". When the 80386DX was introduced in the mid-80s with its 32-bit *Data Bus*, *IBM* decided (much like it did with the *AT*) to create a *Bus* to match this width. *MCA* is 32 bits wide.

Figure 6.6

- *Extended System Configuration Data (ESCD)*

If the *BIOS* were to assign resources to each *PnP* device on every boot, two problems would result:

1. It would take time to do something that it has already done before, each boot, for no purpose. After all, most people change their system hardware relatively infrequently.

2. It is possible that the *BIOS* might not always make the same decision when deciding how to allocate resources, and that might cause changes in the resources even when the hardware remains unchanged.

ESCD is designed to overcome these problems. The *ESCD* area is a special part of the computer's *BIOS' CMOS* memory, where *BIOS* settings are held. This area of memory is used to hold configuration information for the hardware in a computer system. At boot time, the *BIOS* checks this area of memory and if no changes have occurred since the last boot-up, it knows that it does not need to configure anything and skips that portion of the boot process. *ESCD* is also used as a communications link between the *BIOS* and the *Operating System*. Both use the *ESCD* area to read the current status of the hardware and to record changes. *Windows 95* reads the *ESCD* to see if hardware has been changed and react accordingly. *Windows 95* also allows users to override *PnP* resource assignments by manually changing resources in the *Device Manager*. This information is recorded in the *ESCD* area so that the *BIOS* knows about the change at the next boot and does not try to change the assignment back again. The *ESCD* information is stored in a non-volatile *CMOS* memory area, the same way that standard *BIOS* settings are stored. Note that some systems using *Windows 95* can exhibit strange behavior that is caused by incompatibility between how *Windows 95* and the *BIOS* are using *ESCD*. This can cause an "Updating *ESCD*" message to appear each and every time the system is booted, instead of only when the hardware is changed.

CHAPTER VII
MEMORY ARCHITECTURE

- *Memory*

The memory in a computer system is a staging area for the CPU. It is a group of digital circuit elements that can *store* (*write*) data (bits ON and/or OFF), the place where the CPU receives instructions and/or data to process. In an *Operating System (OS)* such as *DOS* or *Windows*, it is the *OS* task to organize a staging area in a prudent and an efficient way for the CPU to exchange instructions and/or data, as the CPU needs it. The CPU uses *Random Access Memory (RAM)* as the main staging area.

One of the main differences among Intel CPUs (for example) is the capability to manage memory:

CPU		*Memory addressing*
8085	8-bit	256
80286	16-bit	65,536
80486	32-bit	4,294,967,295

With the increasing of the CPU speed, the access time of the memory in a computer system can be bottleneck, hence faster memory chips were developed.

An *Operating System* such as *DOS* and *Windows* sometimes uses different types of memory as a supplement to the *RAM* memory. *DOS* and Windows (up to Windows 3.1) use a *segmented memory*; they can handle memory addressing up to the 16-bit range. In a 32-bit CPU, they access *RAM* memory in two 16-bit segments. *Windows* (*3.1*, *NT*, *95* and *98*) use a *virtual memory*. This type of memory can be an area (a space) in the hard disk. *Windows OS* manages the *RAM* and virtual memory as needed. With the help of virtual memory, an *OS*, such as *Windows* can open multiple applications at the same time. For example, a user can open a document, modify a spreadsheet, logon to the Internet and listen to a musical CD disk at the same time. *Windows* (*NT*, *95* and later) also use *flat memory*. This process allows *Windows OS* to directly address the full range of memory (32-bit CPU can address up to 4GB *RAM*).

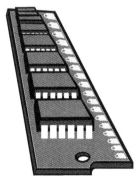

Figure 7.0

A *RAM* chip stores programs when they are running by writing data into *RAM* memory. These chips are referred to as being volatile memory. This is because as soon as the power of the

PC is turned OFF, the information that was stored is no longer there. The data is completely lost unless it is stored somewhere else such as the hard disk or the floppy disk. There are different packages of *RAM* that require specific motherboard types. Some *RAM* chips in a PC might be integrated into the motherboard (chips soldered on *PCB*) that they cannot be changed without changing the motherboard *(Figure 7.1 (a))*. Other *RAM* chips can be inserted into special slots *(Figure 7.1 (b))*. There are also slots or banks that will accept more memory. Older systems use *Dual Inline Package (DIP)* chips *(Figure 7.2)* that would plug into a socket that is mounted on the motherboard. Most modern systems use a 30 or 72 pin memory package called a *Single Inline Memory Module (or SIMM)*. Some newer motherboards accept a 168 pin *Dual Inline Memory Module (or DIMM)*. These memory packages contain several *RAM* chips that are mounted onto it. These modules plug into retaining sockets on the motherboard.

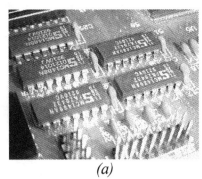 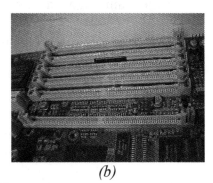

(a)   *(b)*

Figure 7.1

- *Random Access Memory (RAM)*

It is the staging area for the CPU to use as needed. Before a computer can execute an application, perform a function or calculate a math equation; the CPU needs to move instructions or data into an area where it can make decision on what to do. This area is the *Random Access Memory (RAM)*. All data and instructions are stored in *RAM* as bits 1's and 0's. *RAM* is volatile memory, i.e., when the power is OFF, the data stored in memory is lost. There are many types of *RAM* memory:

- o *Static Random Access Memory (SRAM):* It is static because the information stored in this type of memory does not need a constant update or refresh. SRAM stores data as patterns of transistor switches ONs and OFFs to represents the binary digits. This memory is physically bulky and somewhat limited in its capacity. Generally, the memory can store about 256Kbytes per Integrated Circuit (IC). This memory was used in the original early 1980's PC & XT computers and some laptop & notebook computers. Typical access time for SRAM is between 5 and 15 nanoseconds.

- o *Dynamic Random Access Memory (DRAM):* This type of memory uses small capacitors to store data. If a capacitor is charged, the bit of a data is a 1 (one). If a capacitor is discharged, the bit of a data is a 0 (zero). Since DRAM memory uses capacitors instead of transistor switches, it needs to use a constant refresh signal to keep the data in memory. DRAM requires more electrical power than SRAM for refresh signals. DRAM technology allows memory units to be packed with very high density; hence, DRAM ICs can hold very large amounts of data. Typical access time for DRAM is between 50 and 70 nanoseconds.

o *PC100 Synchronous DRAM (SDRAM):* PC100 SDRAM is a SDRAM that states that it meets the PC100 specification from Intel. Intel created the specification to enable RAM manufacturers to make chips that would work with Intel's i440BX processor chipset. The i440BX was designed to achieve a 100 MHz system bus speed. Ideally, PC100 SDRAM would work at the 100 MHz speed. It is reported that PC100 SDRAM improves performance by 10-15% in an Intel Socket 7 system.

o *Extended Data Out (EDO) DRAM:* It is a type of DRAM that is faster than conventional DRAM. Unlike conventional DRAM, which can only access one block of data at a time, EDO RAM can start fetching the next block of memory at the same time that it sends the previous block to the CPU.

o *Burst EDO (BEDO) DRAM*: It is a new type of EDO DRAM that can process four memory addresses in one burst. Unlike SDRAM, however, BEDO DRAM can only stay synchronized with the CPU clock for short periods (bursts). Also, it can't keep up with CPU whose buses run faster than 66 MHz.

o *Rambus DRAM (RDRAM):* It is a type of memory DRAM. RDRAM transfers data at up to 600 MHz. It is already being used in place of VRAM in some graphics accelerator boards.

o *Multi-bank DRAM (MDRAM):* The MDRAM is a relatively new memory technology developed by MoSys Inc. MDRAM utilizes small banks of DRAM (32 KB each) in an array, where each bank has its own I/O port that feeds into a common internal Bus. Because of this design, data can be read or written to multiple banks simultaneously, which makes it much faster than conventional DRAM. Another advantage of MDRAM is that memory can be configured in smaller increments, which can reduce the cost of some components. For example, it is possible to produce MDRAM chips with 2.5 MB, which is what is required by video adapters for 24-bit color at a resolution of 1,024x768. With conventional memory architectures, it is necessary to jump all the way to 4 MB.

o *Synchronous Graphic RAM (SGRAM):* The SGRAM is a type of DRAM used increasingly on video adapters and graphics accelerators. Like SDRAM, the SGRAM can synchronize itself with the CPU bus clock up to speeds of 100 MHz. In addition, SGRAM uses several other techniques, such as masked writes and block writes to increase bandwidth for graphics-intensive functions. Unlike VRAM and WRAM, SGRAM is single-ported. However, it can open two memory pages at once, which simulates the dual-port nature of other video RAM technologies.

o *Video RAM (VRAM):* The VRAM is a special-purpose memory used by video adapters. Unlike the conventional RAM, two different devices can access VRAM simultaneously. This enables the RAMDAC to access the VRAM for screen updates at the same time that the video processor provides new data. VRAM yields better graphics performance but is more expensive than normal RAM.

o *Windows RAM (WRAM):* The *WRAM* is a type of *RAM* developed *by Samsung Electronics* that supports two ports. This enables a video adapter to fetch the contents of memory for display at the same time that new bytes are being pumped into memory. This results in much faster display than is possible with conventional single-port *RAM. WRAM* is similar to *VRAM,* but achieves even faster performance at less cost because it supports addressing of large blocks (windows) of video memory.

o *Fast Page Mode RAM (FPMRAM):* The *FPMRAM* is a type of *DRAM* that allows faster access to data in the same row or page. Page-mode memory works by eliminating the need for a row address if data is located in the row previously accessed. It is sometimes called page mode memory. Newer types of memory, such as *SDRAM* are replacing *FPM RAM.*

o *Single Inline Memory Module (SIMM):* This type of memory was developed because *SRAM* and *DRAM* were taking too much space on a mainboard computer. The idea here is to install the memory on a small circuit board and attach it to the mainboard vertically. Most *SIMM* memory *ICs* are 32-bit wide. For a *Pentium* CPU, where it requires 64-bit wide memory, *SIMM* must be installed in pairs (multiple of two). *SIMMs* are built either with 30 or 72 pins circuit board. SDRAM can deliver data at a maximum speed of about 100 MHz.

o *Single Inline Pinned Package (SIPP):* The *SIPP* is very similar to a *SIMM* except it has pins on the bottom of the module that interface to the motherboard instead of an edge connector like the *SIMMs.*

o *Dual Inline Memory Module (DIMM):* This memory is similar to *SIMM* but is dual sided, i.e., it can hold twice as many ICs as *SIMM. DIMM* is manufactured in 32-bit or 64-bit wide format. *DIMMs* are built either with 72 or 168 pins circuit boards.

Table 7.0 is a list of memory chips that have been used in PC system motherboards.

RAM Chip	Capacity
16K by 1 Bit	Used with IBM PC Type 1 motherboard
64K by 1 Bit	Used with IBM PC Type 2 motherboard, XT Type 1 and 2 motherboards
128K by 1 Bit	Used with IBM AT Type 1 motherboard, often stacked on top of each other and soldered together.
256K by 1 it (also 64K by 4 Bits)	Used in IBM XT Type 2, IBM AT Type 2 motherboards and memory cards
1M by 1 Bit (also 256K by 4 Bits)	Used in 256K to 8M SIMMs
4M by 1 Bit (also 1M by 4 Bits)	Used in 1M to 16M SIMM's
16M by 1 Bit (also 4M by 4 Bits)	Used in 72 Pin SIMM's of 16M and 32M capacity
64M by 1 Bit (also 16M by 4 Bits)	Used for 16M or larger especially in notebooks
256M by 1 Bit (also 64M by 4 Bits)	Newest on the market, allows capacities of 128M or larger

Table 7.0

• *Cache*

The cache memory *(Figures 7.2* and *7.3)*, also known as *high-speed buffer*, is a high-speed memory bank set aside (reserved by the computer) for frequently accessed data or information. Whenever the data is accessed from or committed to the main primary RAM memory, a duplicate image of the data with its address is stored in the cache memory with the associated RAM memory address. The cache memory acts as a go-between the CPU and the primary memory of the computer.

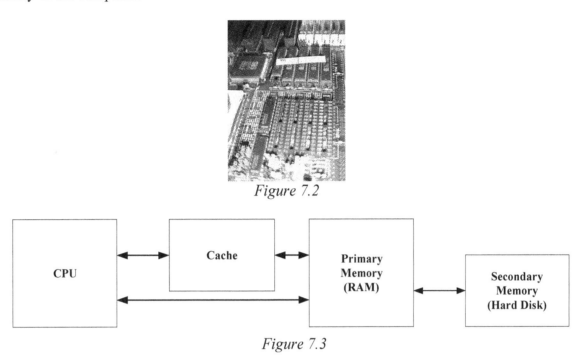

Figure 7.2

```
┌─────────┐      ┌─────────┐      ┌─────────┐      ┌─────────┐
│         │ ◄──► │         │ ◄──► │ Primary │      │Secondary│
│   CPU   │      │  Cache  │      │ Memory  │ ◄──► │ Memory  │
│         │      │         │      │  (RAM)  │      │(Hard Disk)│
└─────────┘ ◄──────────────────► └─────────┘      └─────────┘
```

Figure 7.3

o *Memory cache*: It is a *RAM* that a computer's microprocessor can access more quickly than it can access regular *RAM*. As the microprocessor processes data, it looks first in the cache memory and if it finds the data there (from a previous reading of data), it does not have to do the more time-consuming reading of data from larger memory. *Cache* memory is sometimes described in levels of closeness and accessability to the microprocessor. A *level-1 (L1)* cache is on the same chip as the microprocessor. (For example, the *Motorola's PowerPC 601* CPU has a 32 Kilobytes *level-1 cache* built in it). *Level-2 cache* is usually a separate static *RAM (SRAM)* chip. The main *RAM* is usually a dynamic *RAM (DRAM)* chip. *RAM* itself can be used as a cache of memory for hard disk storage since all of *RAM's* contents come from the hard disk initially when the computer is turned ON and the *Operating System* is loaded onto *RAM* then later an application is started and data or file are accessed. *RAM* can also contain a special area called a *Disk Cache* that contains the data most recently read in from the hard disk.

o *Level 1 (L1) and Level 2 (L2)*: If the computer processor can find the data it needs for its next operation in the cache memory, it will save time compared to accessing it from *DRAM. Level-1 or L1* is a cache memory, usually built onto the microprocessor chip itself. For example, the *Intel MMX* microprocessor comes with 32 KB of *L1. Level-2 or*

55

L2 is also a cache memory on a separate chip (possibly an expansion card) that can be accessed faster than the *DRAM*. A typical *L2* cache memory size is 1 MB.

- o *Disk cache*: It is a mechanism for improving the time it takes to *read from* or *write to* a hard disk. The disk cache is usually included as part of the hard disk. A disk cache can also be a specified portion of *RAM*. The disk cache holds data that has recently been read and, in some cases, adjacent data areas that are likely to be accessed next. Write caching is also provided with some disk caches.

- • *Read-Only Memory (ROM)*

The *ROM* is used to store *Data* permanently for easy and quick retrieval. It is non-volatile memory, i.e., *Data* is stored and saved in memory whether the power is ON or OFF. *ROM* consists of transistor switches that are permanently in the ON or OFF state. Compared to *RAM* memory, data stored in *ROM* memory cannot be changed but accessing the data is extremely fast, also it is expensive to develop and manufacture. There are many types of *ROM* memory:

- o *Programmable Read Only Memory (PROM):* It is a type of *ROM* that initially was manufactured with all transistor circuits in the ON state (logical 0). As the computer is programming the *PROM,* the ON state of a transistor (a *bit*) that needs to be altered is destroyed to the OFF state (logical 1) using a high voltage electrical pulse (between 6.5 *Volts DC* or *VDC* and 28 *VDC*). Typical access time for *EPROM* is between 55 and 250 nanoseconds.

- o *Erasable Programmable Read Only Memory (EPROM):* It is unlike the *PROM, EPROM* is erasable and can be re-programmed, making this type of memory more flexible than *PROM*. A special frequency *ultraviolet (UV)* light shined through a small window on top of the memory wafer can erase the *EPROMs bits*. The exposure of the memory circuit to this light causes the circuits to return to their initial blank state. *EPROM* has to be removed from the circuit board and placed in an *UV* light chamber or box. The original *PC* and *XT* computers used *EPROMs* for *BIOS* data. Typical access time for *EPROM* is between 55 and 250 nanoseconds.

- o *Electrically Erasable Programmable ROM (EEPROM):* It is similar to the *EPROM* in that by sending a special sequence of electrical signals (pulses) to the *EPROM IC,* the memory circuits can be erased. *EEPROM* can still be connected to its circuit board. Typical usage of this type of memory is the *CMOS BIOS* in most current computers.

- o *Complementary Metal-Oxide Semiconductor (CMOS):* It is the memory that usually holds the computer's hardware configuration information such as type of hard drive, amount of *RAM* memory,etc.

- o *Electrically Alterable ROM (EAROM):* It is a memory into which data can be written by a bit or a word, but much more slowly than data readout, also called *Read Mostly Memory*.

- • *Logical Memory Layout of a PC*

With the introduction of the PC, the structure and logical layout of the memory was decided (primarily by *IBM*). Most of that layout is still applied today on PCs with additions of *Extended* and *Expanded* memory.

o *Segment and Offset Memory Addressing*: In order to understand some of the other logical concepts regarding memory, it is useful to understand how the PC refers to memory addresses. For convenience, it is easier to refer to a memory address using a standard or linear address. For example, the *IDE* hard disk *BIOS* may start at memory location C8000H. However, it is not how PC processors refer to memory locations. In x86 CPUs, memory addresses are composed of two parts: the segment address and the offset. These two are added together to produce the "real" address of the memory location, by shifting the segment address one Hex digit to the left (which is the same as multiplying it by 16, since memory addresses are expressed in *Hexadecimal* notation) and then adding the segment offset to it. The address itself is often referred to using the notation *segment:offset*. There are in fact many combinations of segments and offsets that can result in the same linear address. Example, the C8000H again: The standard way to refer to this address is C000:8000; the linear address is C000, shifted one digit to the left to get C0000, and then added 8000 to get C8000. However, C800:0000 results in the same linear address.

o *Memory Layout Overview:* The system memory in the *PC* is divided into several different areas. This design is the legacy of system limitations built into the earliest versions of the *IBM PC* and the versions of *DOS* that ran on them. In 1981, when the *IBM PC* was first released, 1 *MB* was a *lot* of memory. As a result of the design decisions made in the earliest PCs, memory is has the following four basic areas:

- *Conventional Memory Area*: The first 640 KB of the system memory is the conventional memory. This is the area that is available for use by standard *DOS* programs, along with many drivers, memory-resident programs, and most anything else that has to run under standard *DOS*. This is where *DOS* and *DOS* programs conventionally run. It is found at addresses 00000H to 9FFFFH.

- *Upper Memory Area (UMA):* Immediately after (or above) the conventional memory area, this is the upper 384 KB of the first *Megabyte* of system memory. It is reserved for use by system devices and for special uses such as *ROM* shadowing and drivers. It uses addresses A0000H to FFFFFH.

- *High Memory Area (HMA):* This is the first 64 KB (less 16 bytes) of the second Megabyte of system memory. Technically this is the first 64 KB of the *Extended Memory Area,* but it can be accessed when the processor is in real mode, which makes it different from the rest of extended memory. It is usually used for *DOS,* to allow more conventional memory to be preserved. For example, inserting a line *"DOS = HIGH"* in the *DOS* system file *CONFIG.SYS* instructs DOS to load a portion of its own code into the *High Memory Area* instead of the conventional memory. *HMA* occupies addresses 100000H to 10FFEFH.

- *Extended Memory Area (XMS)*: In modern computer systems, the memory that is above the 1 *MB* is used as an *Extended Memory (XMS). Extended Memory* is the most "natural" way to use memory beyond the first *Megabyte,* because it can be addressed directly and efficiently. This memory area is used primarily by all "protected mode" *Operating Systems* (including all versions of *Microsoft Windows*) and programs such as *DOS* games. This is all the memory above the

High Memory Area (above the first *Megabyte* of *RAM*) until the end of system memory. It is used for programs and data when using an operating system running in protected mode, such as any version of *Windows*. Extended memory is found from address 10FFF0H to the last address of system memory. Technically, the *High Memory Area* is part of *Extended Memory Area*. The most commonly used manager is *HIMEM.SYS,* which sets up extended memory according to the extended memory specification *(XMS)*. *XMS* is the standard that PC programs use for accessing extended memory. *HIMEM.SYS* is also used to enable access to the high memory area, which is part of extended memory. Extended Memory is different from *Expanded Memory (EMS)*, which uses bank switching and a page frame in the upper memory area to access memory over 1 *MB*.

- *Expanded Memory (EMS)*: There is, however, an older standard for accessing memory above 1 *MB RAM*, it is called *Expanded Memory*. It uses a protocol called the *Expanded Memory Specification* or *EMS*. *EMS* was originally created to overcome the 1 *MB* addressing limitations of the first generation 8088 and 8086 CPUs. In the mid-80s, when the early systems were still in common use, when running applications such as large spreadsheets in *Lotus 1-2-3*, memory was a problem. To address this problem, a new standard was created by *Lotus, Intel* and *Microsoft* called the *LIM EMS* standard. To use *EMS*, a special adapter board *(PCB)* was added to the PC containing additional memory and hardware switching circuits. The memory on the board was divided into 16 KB logical memory blocks, called pages or banks. The circuitry on the board makes use of a 64 KB block of real memory in the *UMA*, which is called the *EMS Page Frame*. This frame, or window, is normally located at addresses D0000H-DFFFFH, and is capable of holding four 16 KB *EMS* pages. When the contents of a particular part of *EMS* is needed by the PC, it is switched into one of these areas, where it can be accessed by programs supporting the *LIM* specification. After changing the contents of a page, it is swapped out and a new one swapped in. Pages that have been swapped out cannot be seen by the program until they are swapped back in. This concept is called bank-switched memory, and in a way is not all that different from *Virtual Memory*, except here the swapping is not being done to disk but rather to other areas of memory. Note that *EMS* and *XMS* are physically different; an *Expanded Memory* card cannot be used as *Extended Memory,* and *Extended Memory* cannot be used directly as *Expanded Memory*. Example of *EMS* in *MS-DOS* is a driver called *EMM386.EXE* was used in the *CONFIG.SYS* file. The *EMS* emulates the *XMS* by allocating a portion of *XMS*.

- *Upper Memory Block (UMB)*: *UMB* is the upper memory area that is not used by *ROM* or video *RAM* but is generally available for use by other programs. *UMB* typically is not used by regular programs, since it is a small and since a program is usually run in conventional memory. However, it is ideal for loading "memory-resident" programs and drivers (programs for specific hardware devices). To make *UMB* available for drivers requires a driver program to provide access to them. In a *DOS* environment, the driver used is *EMM386.EXE,* which is loaded in the *CONFIG.SYS* file. When this driver is loaded including either the *"RAM"* or *"NOEMS"* parameters, the areas of open memory in the *UMA* are made available for use by drivers and memory-resident programs. A driver in *CONFIG.SYS* can be loaded into the *UMA* by specifying it using *"DEVICEHIGH="* instead of *"DEVICE="*. A program in the *AUTOEXEC.BAT* file can be loaded into the *UMA* using *"LOADHIGH"* or *"LH"* at the front of the command line.

- *ROM Shadowing*: The access time of *ROMs* is usually between 120 and 200 ns, compared to system *RAM* which is typically 50 to 70 ns. Also, system *RAM* is accessed 32-bit at a time, while *ROM* access is 16-bit wide. The result of this is that accesses to the *BIOS* code are very slow relative to accesses to code in the system memory. Since there is *RAM* hiding underneath the *ROM*, most systems have the ability to "mirror" the *ROM* code into this *RAM* to improve performance. This is called *ROM* Shadowing, and is controlled using a set of *BIOS* parameters. There is normally a separate parameter to control the shadowing of the system *BIOS*, the video *BIOS* and adapter *ROM* areas. When shadowing of a region of memory is enabled, at boot time the *BIOS* copies the contents of the *ROM* into the underlying *RAM*, write-protects the *RAM* and then disables the *ROM*. To the system, the shadow *RAM* appears as if it is *ROM*, and it is also write-protected the way *ROM* is. This write-protection is important to remember, because as shadowing of memory addresses that are being used for *RAM* is enabled, the device using it will cease to function when the RAM can no longer be written to (it is locked out by the shadowing). Example, some network cards use parts of the memory region they occupy for both *ROM* and *RAM* functions. Enabling shadowing there will cause the card to hang up due to the write-protection. Similarly, never turn ON shadowing of the regions of memory being used for an *EMS* frame buffer or for *UMBs*.

CHAPTER VIII
STORAGE ARCHITECTURE

- *Storage and External Peripherals*

Originally, in the 1970's, computer information was contained in the system memory and printed to punch cards. To run a program, a computer using a photo-eye light would read the punch cards and detects the holes, interpret that into ASCII codes, and then execute the program. Every line of the program required at least a card; some of the programs had hundreds of cards. The cards had to be in the correct sequence; otherwise the computer would fault the program.

In the late 70's, it was discovered that computer signals could be recorded with a tape cassette recorder. The Read/Write of the cassette recorder was an audio tone. When the tape was played back into the computer, the information was retrieved. This method proved much more efficient and easier than punch cards. There were, however, some limitations of the tape: speed and precision. Tape was slow to store and retrieve programs and data because of the linear nature of tape. Tape devices also were sequential storage devices, i.e., if the information needed was located at the end of the tape, the computer could not accurately fast-forward to the exact position of the tape and retrieve the data. Despite these drawbacks, magnetic tapes were used for data backup and small programs. Today, advanced technology type tape devices are still used for data backup. To overcome these limitations, magnetic disk systems were quickly developed within a year.

- *Media storage*

Media storage refers to various techniques and devices for storing large amounts of data. The earliest storage devices were punched paper cards, which were used as early as 1804 to control silk-weaving looms. Modern mass storage devices include all types of disk drives and tape drives. Mass storage is distinct from memory, which refers to temporary storage areas within the computer. Unlike main memory, mass storage devices retain data even when the computer is turned OFF. The media storage began with the introduction of the Personal Computer, from floppy disk, to hard disk and tape.

Media storage is the capacity of a device to hold and retain informational data. Magnetic disks are the most common form of permanent data storage. Storage capacity began with the hundreds of bytes in the early 80's and today it is the several tens of gigabytes. The main types of mass storage are:

- *Floppy disks (Chapter 9):* Relatively slow and have a small capacity, but they are portable, inexpensive, and universal.

- *Hard disks (Chapter 10):* Very fast and with more capacity than floppy disks, but also more expensive. Some hard disk systems are portable (removable cartridges), but most are not.

- *Tapes backup (Chapter 11):* Relatively inexpensive and can have very large storage capacities, but they do not permit random access of data.

- *Optical disks (Chapter 12):* Unlike floppy and hard disks, which use electromagnetism to encode data, optical disk systems use a laser to read and write data. Optical disks have very large storage capacity, but they are not as fast as hard disks.

Mass storage (also called *auxiliary storage.*) is measured in binary Kilobytes (1,024 bytes), Megabytes (1,024 Kilobytes), Gigabytes (1,024 Megabytes) and Terabytes (1,024 Gigabytes).

- *Redundant Arrays of Inexpensive Disk (RAID)*

Many high-speed computer systems, especially servers, employ a new technology called *Redundant Arrays of Inexpensive Disk* or *RAID*. *RAID* allows for great improvements in both reliability and performance. The idea is to store data on multiple disk drives running in parallel. The primary motivation is reliability. Most *RAID* levels improve performance by allowing multiple accesses to occur simultaneously and by using algorithms that reduce seek time and latency, hence taking advantage of the multiple drives. The exact performance impact depends on the level of *RAID* used; for example, some improve read performance at the expense of write performance.

CHAPTER IX
FLOPPY DISK DRIVE

- *Floppy Disk Drives*

The floppy disk is a soft magnetic storage medium. It is a thin flexible round disk, coated with magnetic oxide and used to store computer data. The floppy disk was used with the CP/M microcomputers in the late 1970's, then with the Personal Computer in the early 1980's and has evolved with the advancement of technology. The word floppy is because the 5¼ inches disk flops if it is waved. Over the years, the size has decreased from 8 inches, to 5¼ inches and then to 3½ inches in diameter (*Figure 9.0*). The capacity has increased from about 120 Kilobytes to 120 Megabytes.

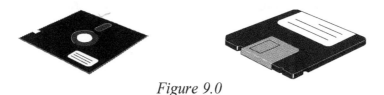

Figure 9.0

Floppy Disk Density is a round plate on which data can be encoded. There are two basic types of disks: **magnetic disks** and optical disks. On magnetic disks, data is encoded as microscopic magnetized **needles** on the disk's surface. Data can be recorded and erased on a magnetic disk many times, similar to a cassette tape.

Unlike the typical hard disks, **floppy disks** (also known as **floppies** or **diskettes**) are portable media. They are inserted and removed from the floppy disk drive of the computer system. Floppy disks are usually slower in accessing the data than hard disk and have less storage capacity, but they are much less expensive. A new disk, called a **blank disk**, has no data on it. Before the data can be stored on a blank disk, however, it must be formatted. Formatting a floppy disk is a *low-level* and a *high-level* formatting.

The hard disk is usually the "data center" of the PC because it is the area where boot area, Operating System software, application software as well as the data files reside. Sometimes, however, the floppy disk is that "data center" for the PC. The first PC (the original *IBM PC*) did not have hard disk drive, instead two floppy drives; all of the PC's data storage was done on floppies. At that time, the floppy disk drive's technology was better than the typical audio cassette tape technology used for data storage. The invention of hard disks relegated floppy disks to the secondary roles of data transfer and software installation. The invention of the CD-ROM and the Internet, combined with the increasingly large size of software files, is threatening even these secondary roles. The floppy disk still persists, basically unchanged for over a decade, in large part because of its universality; the 3½ inch 1.44 MB floppy is present on virtually every PC made since the mid 80's, which makes it still a useful tool. The floppy disk's current role is in these areas:

o *Data Transfer:* The floppy disk is still the most universal means of transferring files from one PC to another. With the use of compression utilities, even moderate-sized files can be shoehorned onto a floppy disk, and anyone can send anyone a disk and feel quite confident that the PC at the other end will be able to read it. The PC 3½ inch floppy is such a standard, in fact, that many Apple and even UNIX machines can read them, making these disks useful for cross-platform transfer.

o *Small File Storage and Backup:* The floppy disk is still used for storing and backing up small amounts of data.

o *Software Installation and Driver Updates:* Many new devices of hardware still use floppies for distributing driver software, and some software still uses floppies. Although this is becoming less common as software grows massive and CD-ROM drives become more universal.

While floppy drives still have a useful role in modern PC systems, there is no denying their reduced importance. Very little attention is paid to floppy "performance" any more. Even choosing the drives' manufacturers or models involves a small fraction of the amount of care and attention required for selecting other components. In essence, the floppy drive today is a commodity item.

The floppy connection from the drive to the mainboard is carried through a flat ribbon cable of 34 wires. These wires include the I/O controller and the Data Bus (*Figure 9.1*).

Floppy Connector Pin Definitions			
Pin Number	Function	Pin Number	Function
1	GND	2	FDHDIN
3	GND	4	Reserved
5	Key	6	FDEDIN
7	GND	8	Index-
9	GND	10	Motor Enable
11	GND	12	Drive Select B-
13	GND	14	Drive Select A-
15	GND	16	Motor Enable
17	GND	18	DIR-
19	GND	20	STEP-
21	GND	22	Write Data-
23	GND	24	Write Gate-
25	GND	26	Track 00-
27	GND	28	Write Protect
29	GND	30	Read Data-
31	GND	32	Side 1 Select
33	GND	34	Diskette

Figure 9.1

Floppy disks come in three basic forms:

- *8-inch (Figure 9.2):* Prior to the Personal Computer era, during the CP/M time, it was the most common. This type of floppy is generally capable of storing between 100KB and 360KB (Kilobytes) of data.

Figure 9.2

- *5¼-inch (Figure 9.3):* Until 1987, this floppy was the common size for Personal Computer. This type of floppy is generally capable of storing between 100KB and 1.2MB (Megabytes) of data. The most common sizes are 360KB and 1.2MB.

Figure 9.3

- *3½-inch (Figure 9.4):* Floppy is something of a misnomer for these disks, as they are encased in a rigid envelope. Despite their small size, microfloppies have a larger storage capacity than the 5¼ inch one: from 400KB to 120 MB of data. The most common sizes for PCs are 720KB, 1.44 MB, 2.88 MB and 120MB.

Figure 9.4

Disk Format	Diameter Size	Max. Capacity
	(inches)	(bytes)
Single Sided, Single Density	**8**	**180 Kilo**
Double Sided, Single Density	**5¼**	**180 Kilo**
Single Sided, Double Density	**5¼**	**360 Kilo**
Double Sided, Double Density	**5¼**	1.2 Mega
Double Sided, Double Density	**3½**	**720 Kilo**
Double Sided, High Density	**3½**	1.44 Mega
Super Disk, High Capacity	**3½**	120 Mega

Table 9.0

The device that spins a disk is called a disk drive. Within each disk drive is one or more *heads* (often called Read/Write Heads) that actually read and write data. Accessing data from a disk is not as fast as accessing data from RAM memory, but floppy disks are cheaper. And unlike RAM, disks hold on to data even when the computer system is turned OFF. Consequently, disks are the storage medium of choice for most types of data. Another storage medium is magnetic tape. But tapes are used only for backup and archiving because they are *sequential-access* devices (to access data in the middle of a tape, the tape drive must pass through all the preceding data).

The new type of floppy disk is called *Super disk* or *High capacity* disk. It can hold over 100 Megabytes. The drives of this floppy can also Read/Write the 1.44 MB floppy disks. The data transfer rates is over 3.5 Megabytes per second (MBps).

Sneakernet is a term used for the method of transferring data information by personally carrying it from one place to another on a floppy disk or other removable medium. The idea is that someone is using their shoes (possibly sneakers) to move data around rather than using the PC network.

- *Floppy Disk interface and configuration*

The floppy disk controller is the piece of hardware responsible for interfacing the floppy disk drives in the PC to the rest of the system. It manages the flow of information over the interface and communicates data read from the floppies to the system processor and memory, and vice-versa.

  o *SCSI Interface Floppy Drives*: This setup is rarely used but sometimes found on some PCs that use specialized industrial applications. The floppy disk is attached to a *SCSI* host adapter using a special cable. This arrangement requires a host adapter that supports floppy disks; not all of them do. Most PCs generally do not use *SCSI* floppy drives, even when *SCSI* interface is used for hard disks and other peripherals. Since the floppy disk is not a primary medium for storage or a primary means of data transfer, it is more cost effective to use a low-cost floppy disk drive instead of an expensive *SCSI* floppy drive.

  o *Floppy Disk Interface*: Regular floppy disks use their own interface, usually called the *floppy disk interface*. This interface is derived from older non-PC designs that were in the 70's systems. The floppy interface is a very simple. Unlike hard disk interfaces where there are compatibility and performance issues galore, with floppy disks it is generally either "it works" or "it does not work". Over time, some other devices have adopted the floppy disk interface, such as tape backup drives.

o *Floppy Disk Controller Implementation:* At one time the floppy disk controller was a dedicated card inserted into an expansion slot in the motherboard. Later, floppy disk controllers were placed onto multifunction controller cards that also provided hard disk interface, serial and parallel ports (I/O card). These cards are commonly found on *ISA*-based and *VLB*-based PCs created from around 1990 to 1994. Since floppy disk controllers have basically not changed in quite some time, it has been possible to standardize and miniaturize them. The latest Pentium-class and later motherboards using *PCI* bus architecture, almost always include floppy disk controllers on the motherboard *PCB*. This support is usually provided through the use of a Super I/O chip that includes support for the floppies, serial/parallel ports and occasionally *IDE/ATA* hard disks.

o *Floppy Disk Controller Speed:* The floppy disk controller included in virtually all new PCs will support every type of standard floppy disk. Older controllers, however, will not work with the newer drives. Generally speaking, the limiting factor is the floppy controller's ability to run at a high enough speed. While the floppy interface is in general very slow--far slower than hard disk interfaces, even at the floppy's top speed--there are in fact different "shades" of slow. The speed required of the controller is directly related to the density of the floppy disk media being used, in particular the bit density per track. Since higher-density floppies record more information in the same space, they require faster data transfer to the drive, to ensure that the data arrives "on time" to be recorded. There are currently three different controller speeds:
  - 250 Kbits per Second: The slowest speed controller, this type will only support the lowest of the floppy densities; this means the double-density 360 KB 5¼ inch and 720 KB 3½ inch drives. This type of controller is obsolete.
  - 500 Kbits per Second: The 500 Kilobits per second controller is found on a large number of PCs that support all disks (360 KB & 1.2 MB 5¼ inch and 720 KB & 1.44 MB 3½ inch drives) except for the 2.88 MB 3½ inch drives.
  - 1 Mbits per Second: The 1 Megabits per second controller supports all of the floppy disk formats on the market (360 KB & 1.2 MB 5¼ inch and 720 KB, 1.44 MB & 2.88 MB 3½ inch drives).

Today, the speed of the floppy controller is actually more important when dealing with floppy interface tape drives. In many cases using the full capacity of the tape drive is dependent upon the floppy controller being fast enough to handle the high data transfer rates required by the tape formats. This is the reason that caused high-capacity floppy formats not to be supported by older, slower controllers.

o *Floppy Disk Controller Resource Usage*: Floppy disks became universal as their resource usages have been all but standardized. In fact, most peripherals reserved the resources that are normally used by the floppy disk controller known in virtually every PC. The floppy disk controller on a standard PC uses the following resources:
  - *Interrupt Request Line (IRQ):* The standard floppy controller *IRQ* is 6. This is occasionally given as an option for tape accelerator cards, because they also use the floppy interface protocol and can be used as replacements for the standard floppy controller. Virtually all other peripherals will avoid *IRQ* 6.

- *DMA Channel*: The standard *Direct Memory Access* (DMA) channel for the floppy controller is 2. Tape accelerator cards offer this is an option while most other peripherals avoid it.

- *I/O Address*: The standard *I/O address* range for the standard floppy controller is 3F0-3F7H. This overlaps with the standard I/O address for the slave device on the primary *IDE* channel. In this case, however, this does not constitute a resource conflict because the overlap is something that has existed for a very long time and is compensated for.

In some cases these resources can be changed manually but they rarely are since they really are the standard settings.

o *Floppy Interface Cable*: The floppy disk interface uses a flat 34-wires gray ribbon cable similar to the standard *IDE* cable *(Figure 9.1)*. There are normally five connectors on the floppy interface cable, although sometimes there are only three. These are grouped into three "sets"; a single connector plus two pairs of two each (for a standard, five-connector cable) or three single connectors. This is what they are used for:

- *Controller Connector*: The single connector on one end of the cable is meant to connect to the floppy disk controller, either on a controller card or the motherboard.

- *Drive A Connectors:* The pair of connectors (or single connector in the case of a three-connector cable) at the opposite end of the cable is intended for the *"A:"* floppy drive.

- *Drive B Connectors*: The pair of connectors (or single connector in the case of a three-connector cable) in the middle of the cable is intended for the *"B:"* floppy drive.

The reason that the standard cable uses *pairs* of connectors for the drives is for compatibility with different types of drives. 3½ inch drives generally use a *pin header connector*, while 5¼ inch drives use a *card edge connector*. Therefore, each position of *"A:"* and *"B:"* has two connectors so that the correct one is available for whatever type of floppy drive being used. Only one of the two connectors in the pair should be used. The three-connector cables are found in some very old systems. They reduce the flexibility of the setup; fortunately these cables can be replaced directly by the five-connector type if necessary. It is noticeable that in a standard floppy interface cable, there is an odd "twist" in the floppy cable, located between the two pairs of connectors intended for the floppy drives. Non-standard floppy interface cables do not have the twist. The reason for the twist is that floppy drives used a *Drive Select (DS)* jumper to configure the drive as either *"A:"* or *"B:"* in the system. Then, special signals are used on the floppy interface to tell the two drives in the system which one the controller is trying to communicate with at any given time. The wires that are crossed-connected via the twist are signals 10 through 16 (seven wires). Of these, wires at pins 11, 13, and 15 are grounds and carry no signal, so there are really four signals that are inverted by the twist. The four signals that are inverted are exactly the ones that control drive selection on the interface. *Table 9.1* shows what happens when the twisted cable is used:

	Wire 10	**Wire 12**	**Wire 14**	**Wire 16**
Controller Signals	Motor Enable A	Drive Select B	Drive Select A	Motor Enable B
Drive Before the Twist Sees	Motor Enable A	Drive Select B	Drive Select A	Motor Enable B
Drive After the Twist Sees	Motor Enable B	Drive Select A	Drive Select B	Motor Enable A

Table 9.1

Since the signals are inverted, the drive, after the twist, responds to commands backwards from the way it should; if it has its drive select jumpers set so that it is an "*A:*" device, it responds to "*B:*" commands, and vice-versa. The reason for that is because it was a big time-saver during the setup when it was common to find two floppy drives in a computer system. Without the twist, using two floppy drives one had to be electrically jumpered as "*A:*" and the other as "*B:*". With the twist, both floppy drives are jumpered as "*B:*" and whichever one is after the twist will appear to the system as "*A:*" because the control lines are inverted (*Table 9.1*). If the computer has only one drive, the connector after the twist is used. Manufacturers, therefore, could arrange to have all of their floppy disk drives configured the same way without having to pull jumpers as the PC are being assembled. In order for this system to work, both drives must be jumpered as "*B:*" drives. Since the floppy cable with the twist is standard, this jumpering scheme has become the standard as well. Virtually all floppy disk drives are pre-jumpered as "*B:*" drives so that they will work with this setup. Some newer BIOS programs have taken things a step further. They include a BIOS parameter that will invert the "*A:*" and "*B:*" signals within the controller itself. When enabled, this lets the user reverse whichever drive is "*A:*" with the one that is "*B:*". Note however that this is not compatible with all operating systems: in particular, both *Windows NT* and *Linux* can malfunction with this swap feature set, which can cause serious problems when trying to install the operating system. The reason this happens is that the swap setting only affects the way the BIOS handles the floppy drive, and confuses operating systems that go directly to the hardware. There are also some manufacturers that use another floppy cable with *two* twists in the floppy cable. The drive placed after the first twist, in the middle of the cable, is "*A:*", as it is with the standard one-twist cable. The drive placed after the second twist is "*B:*". The second twist "reverses" the effect of the first one and makes the connector at the end of the cable operate the same way a drive that appears before the twist in a regular cable does.

o *BIOS Parameters and Issues*: Virtually all PCs support two floppy drives through the standard BIOS setup. The standard BIOS setup screen provides access to the two setup parameters where the system identifies what types of floppy drives are in the system. Support for different drive types is generally based on support from the controller, and the BIOS parameters are arranged so that the drives are selected for the controller to handle. In addition, there are other BIOS parameters that are directly related to the floppy drives:

   ▪ *Boot Sequence:* This parameter controls the order in which the BIOS will try to boot the system.

   ▪ *Floppy Drive Seek*: Some BIOS allows the computer to disable the *"seek"* that the system does to verify the presence of the floppy drive(s) at boot time.

- ▪ *Swap Floppy Drives*: This parameter switches which disk is viewed as *"A:"* and which is *"B:"* by the system.

- o *Floppy Disk Configuration*: Configuring floppy disks that use the standard floppy interface is quite straightforward. Since there can only be one floppy interface, and only two drives on the interface, configuration is simply a matter of making sure each drive has its power connector plugged in securely, and then connecting the floppy cable to each device. The connector on the floppy cable used to attach to each device controls which one appears as *"A:"* and which as *"B:"*. Many problems with floppy drives relate to cables that come loose or are inserted backwards. A drive with the cable in backwards will not function correctly and often, the activity *LED* will come ON and stay ON when the PC is booted.

- o *Floppy Disk Performance*: The floppy disk is the slowest storage device in a PC. Because floppy drives and controllers are standardized, there is nothing any PC user can do to improve hardware performance when it comes to floppies. Traditionally, floppy disks have been accessed using standard BIOS routines and this has contributed to their rather slow performance. Furthermore, their use in a multitasking environment has been quite poor. Using a floppy disk under Windows 3.x platform would seem to slow the entire system to a crawl. Under 32-bit operating systems like Windows 95 and Windows NT, floppy drive access is performed using 32-bit protected mode drivers. This is similar to the way that these operating systems access hard disks for improved performance. The difference in floppy speed under Windows 95, compared to DOS for example, is very noticeable: the floppy disk can be used while performing other tasks.

- o *Floppy Disk Reliability*: Floppy disk drives are generally quite reliable and will last for many years if given reasonable care. An older drive that is dirty and misaligned will cause many more problems than a newer drive that is clean and aligned properly. Failure of floppy disk media is more a matter of when than if. This is true of any device in a computer such as hard disk drive or CD ROM. There are several reasons why floppy disks are generally less reliable than hard disk drives. One is quality: most floppy disk drives are crude affairs assembled in large quantity and sold very cheaply; Floppy disk media is in many cases even worse; competition among manufacturers is frequently based on cost only, since the media is viewed as a commodity item: quality is not an important factor. The nature of floppy disk technology contributes to low reliability as well. While hard disks have the complex task of dealing with a very fast spinning disk and read/write heads floating very near to them on a cushion of air, they do this in a tightly controlled environment. The head assembly is sealed, and the platters are fixed and rigid. Floppy disks use removable media that is not rigid, and both the heads and media are exposed to external contaminants that can damage the media and lead to data loss. Data stored on floppy disk is also subject to loss as a result of stray magnetic fields. In general, floppy disks are reliable only for short-term storage and data transfers. Floppy disks are not recommended to be used for long-term data archives and they are not as a viable backup source for critical data.

- o *Other Devices Using the Floppy Interface*: The floppy disk interface has been used by non-floppy-disk peripherals. However, due to its slow speed, the number of devices that use this interface is limited. Some of the peripherals that use the floppy disk interface are tape drives, by *HP* and *IOmega* (*QIC* and *Travan*). These are inexpensive, *Quarter-Inch Cartridge (QIC)* tape backup drives that make use of the floppy disk interface but do not require high speed. Most of the newer types of drives require the fast 1 Mbits/second controller speed.

- *Floppy Disk Drive Construction and Operation*

While floppy disk drives vary in terms of size and the format of data that they hold, they are all quite similar internally. In terms of construction and operation, floppy drives are similar to hard disk drives, only simpler. Of course, unlike hard disks, floppy disk drives use removable floppy media instead of integrated storage platters.

   o *Read/Write Heads*: The *Read/Write Heads* on the floppy disk are used to convert binary data into electromagnetic pulses, when writing to the disk, or the reverse, when reading. This is similar to what the heads on a hard disk do. There are several important differences between floppy disk and hard disk *Read/Write Heads*. One is that floppy disk heads are larger and much less precise than hard disk heads, because the track density of a floppy disk is much lower than that of a hard disk. The tracks are laid down with much less precision; in general, the technology is more "primitive". Hard disks have a track density of thousands of tracks per inch, while floppy disks have a track density of 135 tracks per inch or less. In terms of technology, floppy disks still use the old ferrite style of head that was used on the oldest hard disks. In essence, this head is an iron core with wire wrapped around it to form a controllable electromagnet. The floppy drive, however, is a contact recording technology. This means that the heads directly contact the disk media, instead of using floating heads that skim over the surface the way hard disks do. Using direct contact results in more reliable data transfer with this more simplistic technology; it is impossible to maintain a consistent floating head gap at any rate when using the flexible media like floppies. Since floppy disks spin at a much slower speed than hard disks, typically 300 to 360 RPM (Rotation per Minute) instead of the 3600 RPM or more of hard disks, they are able to contact the media without causing wearout of the media's magnetic material. Over time, however, some wear does occur, and magnetic oxide and dirt builds up on the heads. This is why floppy disk heads must be periodically cleaned. Contact recording also makes the floppy disk system more sensitive to dirt-induced errors, caused by the media getting scratched or pitted. For this reason, floppy disks are less reliable in the overall measurements than hard disks. The floppy disk also uses a special design that incorporates two erase heads in addition to the *read/write head*. These are called *tunnel-erase heads*. They are positioned behind and to each side of the *read/write head*. Their function is to erase any stray magnetic information that the read/write head might have recorded outside the defined track it is writing. They are necessary to keep each track on the floppy well-defined and separate from the others. Otherwise interference might result between the tracks. Double-Sided floppy disk drives use two heads, one per side, on the drive. On a double-sided disk, the heads contact the media on each side by basically squeezing the media between them when the disk is inserted. The heads for different drives vary slightly based on the drive format and density. Single-sided floppy disk drives basically work the same way except on a one side only.

   o *Head Actuator*: The *Head Actuator* is the device that physically positions the *read/write heads* over the correct track on the surface of the disk. Floppy disks generally contain 80 tracks per side *(Table 9.2)*. The actuator is driven by a *stepper motor*. As the stepper motor turns it moves through various stop positions, and in doing so, moves the heads in and out one or more position. Each one of these positions defines a track on the surface of the disk. Stepper motors were originally used for the actuators for hard disks as well, but were replaced by better coils due to problems with reliability and speed. Since the stepper motor uses pre-defined track placements, thermal expansion in hard disks can cause errors in older hard disks that use stepper motor actuators, when the disk platters expand and move the tracks to a place different than where the heads are expecting them. This is not an issue for floppy disks because of their much lower track density, plus the fact that thermal expansion is not nearly as big of an issue for floppies.

Floppy Disk tracks		
Size (inch)	No. of tracks	Capacity (bytes)
5¼	40	360 K
5¼	80	1.2 M
3½	80	720 K
3½	80	1.44 M
3½	80	2.88 M

Table 9.2

Over time, however, a floppy disk can develop difficulties if the track positioning of the actuator drifts from what is normal. This is called a head alignment problem. When the heads become misaligned, disks will work if formatted, written and then read in the same drive, but not if used in an aligned drive. This is because the formatting of the floppy is what defines where the data is placed. Misalignment can be resolved by having the heads on the floppy disk realigned. If the floppy drive is misaligned and since the realignment labor costs more than a new drive, it is recommended to purchase a drive.

The *head actuators* on a floppy disk are very slow, compared to hard disks, which makes their *seek time* much higher. While a hard disk's actuator can move from the innermost to outermost tracks (full-stroke seek) in about 20 milliseconds, a floppy disk will typically take 10 times that amount of time or more. This is one reason why floppy disks are much slower than hard disks.

o *Spindle Motor*: The spindle motor on the floppy is what spins the floppy disk when it is in the drive. When the disk is inserted, clamps come down on the middle of the disk to physically grasp it. These clamps are attached to the spindle motor, which turns the disk as it spins. The speed of the spindle motor (*Table 9.3*) depends on the type of floppy drive:

	360 KB 5¼"	1.2 MB 5¼"	720 KB 3½"	1.44 MB 3½"	2.88 MB 3½"
Spindle Speed	300 RPM	360 RPM	300 RPM	300 RPM	300 RPM

Table 9.3

The very slow spindle speeds used for driving floppy disks is another major reason their performance is so poor compared to other media, since the spindle speed affects both latency and data transfer rate It is this slow speed however that allows the heads to ride contacting the surface of the media without causing the floppy disk's magnetic coating to wear right off. The spindle motor on a floppy also uses very little power and generates very little heat due to its simple needs.

o *Disk Change Sensor*: Modern floppy drives incorporate a special sensor and a signal on the floppy cable that work in conjunction to tell the floppy controller when a disk is ejected and a new one inserted. This signal is used for performance reasons such as keeping track of whether or not the disk is changed. Knowing this saves the system from having to constantly re-examine the disk each time the floppy is accessed to see what is there. Otherwise each time the disk is referenced, the disk's structures would have to be re-examined, causing a great performance penalty.

o *Logic Board*: The floppy disk contains an integrated logic board that acts as the drive controller. The logic board contains the electronics that control the *read/write heads*, the *spindle motor*, *head actuator* and other components. The circuits on this board also communicate to the floppy disk controller over the floppy disk interface. SCSI floppy disks include a SCSI interface chip on the logic board to communicate with the SCSI interface.

CHAPTER X
HARD DISK DRIVE

- ### *Fixed Hard Disk*

A hard disk is part of a unit, often called a *"Disk Drive"*, *"Hard Drive"* or *"Hard Disk Drive"* or *HDD* (*Figure 10.0*) that stores and provides relatively quick access to large amounts of data on an electromagnetically charged surface or set of surfaces. *Hard Disk Drives* (*HDD*) for PCs generally have seek times of about 12 milliseconds or less. Many disk drives improve their performance through a technique called *caching (Disk cache, Chapter 7)*. Hard disk drives are sometimes called *Winchester drives,* Winchester being the name of one of the first popular hard disk drive technologies developed by *IBM* in 1973. Today's computers typically come with a hard disk that contains several billion bytes (Gigabytes) of storage space.

Figure 10.0

A hard disk is really a set of stacked *"disks"* each of which, like phonograph records, has data recorded electromagnetically in concentric circles or *"tracks"* on the disk. A *"head"* (something like a phonograph arm but in a relatively fixed position) *writes* or *reads* the information on the tracks. Two heads, one on each side of a disk, read or write the data as the disk spins. Each read or write operation requires that data be located, which is an operation called a *"seek"*. Data already in a disk cache, however, will be located more quickly. A hard disk/drive unit comes with a set rotation speed varying from 4500 to 7200 rpm. Disk access time is measured in milliseconds.

Although the physical location can be identified with cylinder, track, and sector locations, these are actually mapped to a *Logical Block Address (LBA)* that works with the larger address range on today's hard disks.

The hard disk is also a magnetic storage medium that began with the Personal Computer in the early 1980's. It is a magnetic disk on which a computer system data can be stored. The term *hard* is used to distinguish it from a soft, or *floppy*, disk. A hard disk can hold more data, for example, it can store anywhere from 10 Megabytes to more 10 Gigabytes whereas most floppies have a maximum storage capacity of 1.44 Megabytes. Hard disks are also 10 to 100 times faster than floppy disks. A single hard disk usually consists of several platters. Each platter requires two *Read/Write Heads*, one for each side. All the read/write heads are attached to a single access arm so that they cannot move independently. Each platter has the same number of tracks, and a *track* location that cuts across all platters is called a *cylinder*. For example, a typical 121

73

Megabytes hard disk for a Personal Computer might have two platters (four sides), 15 *Heads*, 17 *Sectors* and 917 *Cylinders*. Typical access time for fixed hard disk is between 6 and 18 *milliseconds* (more than 200 times slower than average DRAM access time).

- **Hard Disk Interface**

The interface that the hard disk uses to connect to the rest of the PC is in many ways as important as the characteristics of the hard disk itself. The interface is the communication channel over which all the data flows that is read from or written to the hard disk. The interface can be a major limiting factor in system performance. It also has important implications for system configuration, upgradability and compatibility. Over time, several different standards have evolved to control how hard disks are connected to the other major system components used in the PC. These have tended to build upon one another, and often use confusing and overlapping terminology. The result has been a great deal of confusion surrounding the entire subject.

Every hard disk interface communicates with the PC over one of the system's *I/O Buses*. This usually means the *PCI*, *VLB* or *ISA Bus*. Logically, the hard disk interface is one device on the system I/O bus, which is connected to the memory, processor and other components of the system. The choice of bus type has a great impact on the features and performance of the interface. Higher-performance interfaces, including the faster transfer modes of both *IDE/ATA* and *SCSI*, require an interface over a local *Bus*, which means either *PCI* or *VLB*. The *ISA Bus* is too slow for high-speed transfers, and is still found only on older PC systems. The speed of the system Bus is also important. The standard speed for the *PCI* Bus on Pentium-class systems is 25, 30 or 33 MHz. The faster the system's Bus, the faster the hard disk's access. Some older PC systems use a dedicated hard disk interface card that goes into a system Bus slot and then connects to the drives internally. Other newer PCs that use *PCI* Bus have the ports for two *IDE/ATA* channels built into the motherboard itself. In practical terms, there is no real difference except for the cost savings associated with not needing to put a separate hard disk interface card into the PC when it is built into the motherboard.

There is an interface controller between the CPU and the media storage device. The interface controller has gone through many technologies over the years, from the slow transfer rate of 125 Kilobits per second up to more than 33 Megabits per seconds. Some of the interface controllers are:

- o **Modified Frequency Modulation (MFM):** The *MFM* interface controller was developed to resolve the minimal storage and speed of the floppy disk. The 8-bit *MFM* was the initial fixed mass storage interface for the personal computer. Mass storage was available for up to 5 megabytes. The 8-bit MFM data burst transfer was 125 Kilobits per second. Then the 16-bit *MFM* interface was introduced as an enhancement with data burst transfer up to 400 Kilobits per second.

- o **Run Length Limited (RLL):** Shortly after the introduction of the *MFM* technology, the *Run Length Limited (RLL)* interface controller was introduced to compete with it. The 8-bit *RLL* data burst transfer was 175 Kilobits per second. Then the 16-bit *RLL* interface was introduced as an enhancement with data burst transfer up to 550 Kilobits per second.

- o **ST-506 / ST-412 Interface:** The original hard disk interface used on the original *XT* was developed by *Seagate Technologies*. It was used on their first drives and became the first standard for use in PCs. The name *ST-506/ST-412* comes from the first two model numbers of hard disks made by Seagate. In common parlance, this older interface is sometimes called "*MFM*" or "*RLL*". These actually refer to the encoding method used for storing data on the disk. Both encoding methods were used on these early drives. In fact, *RLL* is used on some *IDE* and *SCSI* drives also, making it a poor name for referring to drives of one particular interface. The *ST-506/ST-412* interface differs from the *IDE/ATA* and *SCSI* standards in one very important respect: the hard disks were "dumb", meaning there was no built in logic board as in newer drives. All of the smarts resided in the controller card that plugged into the PC. This caused a host of problems relating to compatibility, data integrity and speed, because the raw data from the *read/write heads* was traveling over a cable between the controller and the drive. This interface also required a lot more work on the part of the user, because while a newer drive ships with an integrated controller card built into the drive that is optimized for that drive, these older ones did not have this, and therefore the person setting up the drive had to program the interleave ratios and other factors into the drive to achieve maximum performance. By today's standards, this interface and the drives that use it are small in capacity (although sometimes enormous in physical size), slow, cumbersome, error-prone and completely obsolete. This interface in older PC systems had two flat ribbon cables: 20-pins data and 34-pins control signals.

o **Enhanced Small Device Interface (ESDI):** The *ESDI* was developed in the mid 1980's and was eventually codified as an *ANSI* standard. *ESDI* was a much needed interface controller due to the limitation and lack of speed of the *MFM* and *RLL* interfaces. The data burst transfer rate is 24 Megabits per second. The *MFM, RLL* and *ESDI* used similar cables: a 20-pins data cable and 34-pins controller cable. *ESDI* was the first attempt at improving the original *ST-506/ST-412, MFM* and *RLL* hard disk interface by moving some drive controller functions from the controller card to the hard disk, increasing the data throughput rate and making other performance enhancements. *ESDI* became obsolete itself very quickly due primarily to *IDE/ATA* interface, which was simpler, cheaper and offered higher performance.

o **Integrated Device Electronics (IDE):** There are three main types of *IDE* interfaces that are available, with the differences based on three different standards (These three interfaces are not compatible with each other):

- **AT Attachment (ATA) IDE (16-bit ISA):** *Compaq Corporation, CDC* and *Western Digital Corporation* created what was called the first *ATA* type *IDE* interface drive and were the first to establish the 40-pin *IDE* connector pin-out (*Table 10.0*).

Hard Drive Connector			
IDE Connector Pin Definitions			
Pin Number	Function	Pin Number	Function
1	Reset IDE	2	GND
3	Host Data 7	4	Host Data 8
5	Host Data 6	6	Host Data 9
7	Host Data 5	8	Host Data 10
9	Host Data 4	10	Host Data 11
11	Host Data 3	12	Host Data 12
13	Host Data 2	14	Host Data 13
15	Host Data 1	16	Host Data 14
17	Host Data 0	18	Host Data 15
19	GND	20	Key
21	DRQ3	22	GND
23	I/O Write-	24	GND
25	I/O Read-	26	GND
27	IOCHRDY	28	BALE
29	DACK3-	30	GND
31	IRQ14	32	IOCS16-
33	Addr1	34	GND
35	Addr0	36	Addr2
37	Chip Select 0	38	Chip Select 1-
39	Activity	40	GND

Table 10.0

In 1989 the *CAM ATA (Common Access Method AT Attachment)* interface committee introduced the first working document of the *ATA* interface. Before this, there were many different proprietary *IDE* interfaces that were difficult to integrate into a dual drive setup with newer drives. *ATA-1* was approved in 1994 and *ATA-2* (also called *Enhanced IDE*) was approved in 1995. Some of the functions that are defined by the *ATA* specification are:

- Dual-Drive Configurations

- *AT I/O* Connector

- *ATA I/O* Cable

- *ATA* Signals

- *ATA* Command Set

There are three main *ATA IDE* drive categories:

76

- *Non-Intelligent ATA-IDE drives:* These drives do not respond to the Identify command. They also do not support sector translation, in which the physical parameters could be altered to appear as any set of logical cylinders, heads, and sectors.

- *Intelligent ATA-IDE drives:* These drives do support the Identify command and also sector translation.

- *Intelligent Zoned Recording ATA-IDE drives:* This is the most sophisticated *IDE* drive. With *Zoned Recording*, the drive has a variable number of sectors per track in several zones across the surface of the drive. Because the *PC BIOS* can handle only a fixed number of sectors on all tracks, these drives always must run in translation mode. Because these drives are always in translation mode, the factory-set setup such as interleave and skew factors cannot be altered or the factory defect information be deleted.

The *IDE* data burst transfer is between 3.3 and 33.3 Megabits per second (*Mbps*). It can support mass storage devices of up to 528 Megabytes (*MB*).

- o *Enhanced IDE (EIDE or ATA-2):* The *Enhanced IDE* (*Expanded IDE or EIDE)* was a newer version of the *IDE* mass storage device interface standard developed by *Western Digital Corporation. EIDE* was a standard electronic interface between the computer and its mass storage drives. *EIDE* also provided faster access to the hard drive, support for Direct Memory Access (DMA), and support for additional drives, including *CD-ROM* and tape backup devices through the *AT Attachment Packet Interface (ATAPI)*. When updating a computer with a larger hard drive (or other drives), an *EIDE* "controller" can be added to the computer in one of its card slots. It supports data rates between 4 and 16.6 Megabits per second (*Mbps*), about three to four times faster than the old *IDE* standard. In addition, it can support mass storage devices of up to 8.4 Gigabytes (*GB*), whereas the old standard was limited to 528 Megabytes (*MB*). To access larger than the 528 Mbytes drives, *EIDE* (or the *BIOS* with it) used a 28-bit *Logical Block Address (LBA)* to specify the actual cylinder, head, and sector location of data on the disk. The 28-bit of the *LBA* provided enough information to specify unique sectors for a device up to 8.4 GB in size. *EIDE* was adopted as a standard by *ANSI* in 1994. *ANSI* calls it *Advanced Technology Attachment-2*. Because of its lower cost, *EIDE* has replaced SCSI in many areas. *EIDE* is sometimes referred to as *Fast ATA* or *Fast IDE,* which is essentially the same standard, developed and promoted by Seagate Technologies. It is also sometimes called *ATA-2*. There are four *EIDE* modes defined. The most common is *Mode 4*, which supports transfer rates of 16.6 Megabytes per second (*MBps*). There is also a new mode, called *ATA-3* or *Ultra ATA*, that supports transfer rates of 33 Megabytes per second (*MBps*). The most important additions are performance, enhancing features such as fast *PIO* and *DMA* modes. *ATA-2* also features improvements in the Identify command allowing a drive to tell the software exactly what its characteristics are; this is essential for both *Plug and Play (PnP)* and compatibility with future revisions of the standard. There are four main areas where *ATA-2* has improved the original *ATA/EIDE* interface:

  - ▪ *Increased maximum drive capacity:* This is done through an Enhanced *BIOS*, which makes it possible to use hard disks exceeding the 504 MB barrier. The origin of this limit is the disk geometry (cylinders, heads, sectors) supported by the combination of an *IDE* drive and the *BIOS'* software interface. Both *IDE* and the *BIOS* are capable of supporting large size disks, but their combined limitations conspire to restrict the useful capacity to 504 MB.

- *Faster data transfer: ATA-2/EIDE* defines several high-performance modes for transferring data to and from the drive. These faster modes are the main part of the new specifications and were the main reason they were initially developed. *Table 10.1* shows the *PIO* modes, with their respective data transfer rates:

PIO Mode	Cycle Time (ns)	Transfer Rate (M/sec)	Specifications
0	600	3.3	ATA
1	383	5.2	ATA
2	240	8.3	ATA
3	180	11.1	ATA-2
4	120	16.6	ATA-2

Table 10.1

- *Secondary two-device channel:* The *ATA/IDE* specification was ironed out after many companies were already making and selling drives, many older *IDE* drives have problems in dual drive installations, especially when the drives are from different manufacturers. The manufacturers that followed the specification had no problem in dual drive installations. Most *IDE* drives came in three configurations:

  - Single-drive (master)

  - Master (dual-drive)

  - Slave (dual-drive)

- *ATAPI (ATA Program Interface): ATAPI* is a standard for CD-ROM's and tape backup drives that plug into an ordinary *ATA/IDE* connector. The principal advantage of *ATAPI* hardware is that it is cheap and works on the existing *IDE* adapter. For *CD-ROMs*, it has a somewhat lower CPU usage compared to proprietary adapters, but there is no performance gain otherwise.

o *Small Computer System Interface (SCSI):* The *SCSI* is not a disk interface, but a systems-level interface. *SCSI* is not a type of controller, but a Bus that supports as many as eight devices. One of these devices, the host adapter, functions as the gateway between the *SCSI* Bus and the PC system Bus. This interface has its roots in *SASI* (*Shugart Associates System Interface*). The *SCSI* Bus itself does not communicate directly with devices such as hard disk; instead, it communicates with the controller that is built into the drive. It is a fast and flexible interface. The *SCSI* is a set of evolving *ANSI* standard electronic interfaces that allow personal computers to communicate with peripheral hardware such as disk drives, tape drives, *CD-ROM* drives, printers, and scanners faster and more flexibly than previous interfaces. *SCSI* ports are built into most personal computers today and supported by all major operating systems. The data burst transfer of *SCSI* is between 5 and 80 Megabytes per second (*MBps*). In addition to faster data rates, SCSI is more flexible than earlier parallel data transfer interfaces. The *SCSI* was designed for multiple devices that are connected in a daisy-chained form (one device connected to another). Each device must have a unique ID number (*Address*). The devices at the beginning and the end of the daisy-chained must be terminated with a resistor. The original *SCSI* uses a single 50-pin cable.

The latest *SCSI* standard, *Ultra-2 SCSI* for a 16-bit Bus which uses a 40 MHz clock rate to get maximum data transfer rates up to 80 Megabytes per second (*MBps*). *Ultra-2 SCSI* allows up to 7 or 15 devices (depending on the Bus width) to be connected to a single SCSI port in daisy-chain fashion. This allows one circuit board or card to accommodate all the peripherals, rather than having a separate card for each device, making it an ideal interface for use with portable and notebook computers. A single host adapter, in the form of a *PC* Card, can serve as a *SCSI* interface for a "laptop", freeing up the parallel and serial ports for use with an external modem and printer while allowing other devices to be used in addition. Although not all devices support all levels of *SCSI*, the evolving *SCSI* standards are generally backwards-compatible. That is, an older device can be attached to a newer computer with support for a newer standard, the older device will work at the older and slower data rate. The original *SCSI*, now known as *SCSI-1*, evolved into *SCSI-2*, known as "*SCSI*" as it became widely supported. *SCSI-3* consists of a set of primary commands and additional specialized command sets to meet the needs of specific device types. The collection of *SCSI-3* command sets is used not only for the *SCSI-3* parallel interface but also for additional parallel and serial protocols, including *Fibre* Channel, Serial Bus Protocol and the *Serial Storage Protocol (SSP)*. It provides a longer possible cabling distance (up to 12 meters or about 36 feet) by using *Low Voltage Differential (LVD)* signaling. Earlier forms of *SCSI* use a single wire that ends in a terminator with a ground. *Ultra-2 SCSI* sends the signal over two wires with the data represented as the difference in voltage between the two wires. This allows support for longer cables. A low voltage differential reduces power requirements and manufacturing costs. *Table 10.2* shows a summary of existing *SCSI* standards.

Name	Maximum Cable Length (ft)	Maximum Speed (Mbps)	Maximum Number of Devices
SCSI-1	20	5	8
SCSI-2	20	5 - 10	8 or 16
Fast SCSI-2	10	10 - 20	8
Wide SCSI-2	10	20	16
Fast Wide SCSI-2	10	20	16
Ultra SCSI-3, 8-bit	5	20	8
Ultra SCSI-3, 16-bit	5	40	16
Ultra-2 SCSI	40	40	8
Wide Ultra-2 SCSI	40	80	16

Table 10.2

- ### *Hard Disk Configuration*

The configuration of a hard disk begins with the *"type"* of the *IDE* device. In the early years of the PC, there were few different types of hard disks, and there were far less sophisticated *BIOS* setup programs. There was no *autodetection* for hard disks, and no way to manually enter the parameters for the hard disk. The *"type"* was selected by a number (usually from 1 to about 45) of the hard drive from a predefined table that was entered into the *BIOS* software program. Different machines would have different tables, and newer PCs would have more drives in their tables than older PCs did. If a new drive in an older PC was to be added and had no entry in the *BIOS* that matched, the "best fit" entry had to be used which sometimes cause losing some of the drive's capacity. New *BIOS* programs are not restricted to the use of the entries in the fixed *"disk type"* table, although the table of fixed entries still persists. In today's BIOS, a system would normally have the following options for each device's *"Type"*:

  o ***Predefined Types (1 thru 45, 1 thru 46, or 1 thru 47):*** This is the predefined fixed table. Even on new machines this table generally contains entries for small capacity, old drives (such as 10-100 MB). It is recommended to avoid the use of this table because new hard drives have built in setups that *BIOS* can detect.

  o ***User:*** This option lets the user manually specify the parameters for the drive. Not recommended unless the user really knows what to do or there is a specific reason to do this. *"User"* is normally what the BIOS will set the drive to in a manual *autodetect*. Instead of *"User"*, some PC systems use the last number in the table for user settings. For example, if entries 1 through 46 are predefined drives, the BIOS may call the *"User"* setting *"Type 47"*.

  o ***Auto:*** This setting activates dynamic *IDE autodetection* for the device. The hard drive will be autodetected by the BIOS each time the system boots.

  o ***CD-ROM:*** Some new PC systems support this entry, to let the BIOS know that the computer is using a CD-ROM drive as an *IDE* device.

  o ***Disabled / None:*** Use this option to let the BIOS know that there is no drive at all in this particular *IDE* device position.

  Some BIOS programs implement manual autodetection of *IDE* devices using the *"Type"* setting. By pressing {Enter} key from the keyboard, the BIOS will autodetect the device, set the type to *"User"* then set the other numbers and options for the computer. Most *BIOS* programs however have a dedicated menu entry for autodetecting all *IDE* devices. If a selection is made for *"Type"* other than *"User"*, the BIOS will lock the *"Size"*, *"Cylinders"*, *"Heads"*, *"Sectors"*, *"Write Precompensation"* and *"Landing Zone"* settings, since these will be determined either by reading the fixed table, or by dynamic autodetection (*"Auto"* selection). CD-ROM drives do not use these physical geometry parameters since their construction is totally different.

  If the computer system supports the *"Auto"* setting, it is best to use it. This will ensure that the system is always set up correctly. It is recommended to disable any devices that are not used and use *"CD-ROM"* for *IDE* CD-ROM drives. Some *BIOS* programs do not have a *"Type"* entry for CD-ROMs, but will autodetect a CD-ROM at boot time. In this case, it is best to set that device to *"Auto"*.

o **Size:** This setting indicates the "*size*" of the drive, normally in decimal Megabytes. Since the value is not normally entered. It is typically calculated based on the following equation:

$$Size = \frac{Heads * Cylinders * Sectors * 512}{10^6} \quad (Eq.\ 10.0)$$

Note: The 512 is because PCs use 512-Bytes sectors; the 10^6 value converts the number into decimal Megabytes. The number of heads, cylinders and sectors in this formula is equal to whatever the settings are in the *BIOS*.

o **Cylinders:** They are the number of cylinders on each side of each platter in the disk. For older drives, this was the number of *physical* cylinders the disk uses.

The generic term used to refer to the way the disk structures its data into platters, tracks and sectors, is its *geometry*. In the early days this was a relatively simple concept: the disk had a certain number of heads, tracks per surface, and sectors per track. These were entered into the *BIOS* set up so the PC knew how to access the drive. With newer drives the situation is more complicated. The simplistic limits placed in the older *BIOS* have persisted to this day, but the disks themselves have moved on to more complicated ways of storing data, and much larger capacities. For newer disks, it is the *logical* number of cylinders that the drive specifies for use in the *BIOS* setup:

- **Physical Geometry:** The *Physical Geometry* of a hard disk is the actual physical number of heads, cylinders and sectors used by the disk. On older disks this was the only type of geometry used. The original setup parameters in the system *BIOS* are designed to support the geometries of those older drives, in particular the fact that every track has the same number of sectors. All newer drives that use zoned bit recording must hide the internal physical geometry from the rest of the system, because the *BIOS* can only handle one number for sectors per track. These drives use logical geometry figures, with the physical geometry hidden behind routines inside the drive controller.

- **Logical Geometry:** The *logical geometry* parameters that the hard disk manufacturer has specified for the drive is when a computer performs a drive parameter autodetection in *BIOS* setup or look in a new hard disk's setup manual to see what the drive parameters are. Since newer drives use zoned bit recording and hence have ten or more values for sectors per track depending on which region of the disk is being examined, it is not possible to set up the disk in the *BIOS* using the physical geometry. Also, the *BIOS* has a limit of 63 sectors per track, and all newer hard disks *average* more than 100 sectors per track, so even without zoned bit recording, there would be a problem. To get around this issue, the *BIOS* is given "bogus" parameters that give the approximate capacity of the disk, and the hard disk controller is given intelligence so that it can do automatic translation between the logical and physical geometry. Virtually all-new hard disks use a logical geometry with 16 heads and 63 sectors, since these are the largest values allowed by the *BIOS*.

The fact that both geometries equate to the same number of total sectors is *not* a coincidence. The purpose of logical geometry is to enable access to the entire disk using terms that the *BIOS* can handle. The logical geometry could theoretically end up with a smaller number of sectors than the physical, but this would mean wasted space on the disk. It cannot specify *more* sectors than physically exist. The translation between logical and physical geometry is the lowest level of translation that occurs when using a new hard disk. It is different from BIOS geometry translation, which occurs at a higher level.

While the use of logical hard disk geometry gets around the problem that physical hard disk geometries cannot be properly expressed using standard *BIOS* settings, they do not go far enough. In most cases, higher levels of translation are needed as well because other problems relating to old design decisions make it impossible for even logical geometry to be used with modern large hard disks. These are the infamous *BIOS* capacity barriers such as the 504 MB limit on standard *IDE/ATA* hard disks, and other similar issues. In order to get around these barriers, another layer of translation is often applied on top of the geometry translation that occurs inside the hard disk. This translation is performed by the *BIOS*. There are many issues involved in *BIOS*-level translation.

- o **Heads:** They are the number of *read/write* heads the disk uses. For older drives, this is the number of physical heads the disk uses. For newer disks, it is the logical number of heads that the drive specifies for use in the *BIOS* setup, usually 16. For new drives supported using *BIOS* geometry translation, the *BIOS* increases the number of heads by the same number it uses to reduce the number of cylinders, so that the number of cylinders is less than the *BIOS* limit of 1,024. For example, a Western Digital Caviar 33100 3.1 GB drive has nominal parameters (*logical geometry*) of 6,136 *cylinders*, 16 *heads* and 63 *sectors*. When it is setup (*autodetect*), the *BIOS* will record 767 *cylinders*, 128 *heads* and 63 *sectors*. The drive does not really have 128 heads, but this is the way that the BIOS gets around the 504 MB restriction (*Chapter 19*).

- o **Sectors:** This setting is the number of sectors on each track (cylinder and head combination). A sector contains 512 bytes and is the smallest unit of data normally referenced on a hard disk. For older drives that use the same number of sectors per track, this is the number of physical sectors on each track; the most common number is 17. Newer drives use zoned bit recording to place a different number of sectors on tracks in different parts of the disk. The *BIOS* only allows a single number for sectors per track, these drives use a logical geometry, often specifying 63 sectors per track, the largest number that will fit in this field. The drive translates internally to the correct sector numbers.

- o **Write Precompensation:** Old disks that use the same number of sectors for every track sometimes required an adjustment to be made while writing data onto the disk, beginning at a certain track number, and this setting was that value. Modern *IDE/ATA* and *SCSI* drives have built-in intelligent controllers that take care of those types of adjustments automatically. This setting should normally be set to -1, 0 or 65535 (the largest value it can support) or whatever value the *autodetection* sets. These values tell the *BIOS* that "there is no write *precompensation* value for this drive". The number itself is ignored by the drive in any event.

- o **Landing Zone:** This setting specifies the cylinder to which the BIOS should send the heads of the hard disk when the machine is to be turned OFF. This is where the heads will "land" when they spin down. Modern drives automatically park the heads in a special area that contains no data when the power is turned OFF. Therefore this setting is meaningless and is typically ignored. Most BIOS set this value to be the largest cylinder number of the logical geometry specified for the disk when it is auto-detected. So if the drive has 6,136 logical cylinders, the landing zone will be set to 6,135. In any event a modern *IDE* drive will ignore this setting and auto-park by itself.

o **Translation Mode:** This setting specifies the translation and/or addressing mode for the drive. These special modes are used to enable the *BIOS* (and the operating system) to handle large hard drives and overcome hard disk capacity barriers, especially the *504 Binary MB* barrier (*Chapter 19*). Some of the options for this setting are:

▪ **Normal or CHS Mode:** This mode is sometimes called "*CHS*" mode, for "*Cylinder, Head, Sector*", the three geometry specifications for a hard disk. This is the "standard" mode with no special translation or addressing. It is used for regular IDE/ATA hard disks that are smaller than 504 MB; more precisely, it should be used for any hard disk that has 1,024 or fewer cylinders *and* 16 or fewer heads.

▪ **Logical Block Addressing (LBA):** *LBA* is a method used with *SCSI* and *IDE* disk drives. *LBA* is a technique that allows a computer to address a hard disk larger than 528 megabytes. *LBA* is used to serialize the sectors with an integer number (0, 1, 2, …) up to the total number of sectors on the disk instead of referring to locations by passing to the disk a cylinder, head and sector number (*CHS addressing*). It is one of the defining features of *EIDE*, a hard disk interface to the computer Bus or data paths. *LBA* is currently the standard for addressing large hard disks, and is recommended for hard disks that are not small enough to be used under "*Normal*" mode. In *LBA* mode, the *autodetect* program will still translate the drive parameters so that the number of cylinders is less than 1,024, the *BIOS* limit. However, accesses to the disk will be based on the integer sector number. *LBA* is a 28-bit value that maps to a specific cylinder-head-sector address on the disk. The 28-bit allows sufficient variation to specify addresses on a hard disk up to 8.4 Gigabytes in data storage capacity.

▪ **Large:** This mode is also called "*Extended CHS*" or "*ECHS*" mode. This mode uses translation to ensure that the number of cylinders is less than 1,024. However, unlike *LBA*, it does not number the sectors linearly, it refers to the disk using the translated cylinder, head and sector values. This is a valid way to deal with larger hard disks, however it is very rarely used and is considered non-standard.

▪ **Auto:** Some *BIOS* will automatically detect and set the hard disk mode at boot time. In general, some BIOS have the ability to dynamically autodetect all drive settings at boot time. However, even if this overall boot-time autodetection is not used, this specific mode autodetection can be used if the BIOS supports it. The BIOS autodetection program will normally take care of making the appropriate mode selection for the computer (either if in "Auto" setting or the drive is autodetected).

Table 10.3 shows an example the difference between the physical and logical geometry for a 3.8 GB Hard Disk:

Specification	Physical Geometry	Logical Geometry
Read/Write Heads	6	16
Cylinders (Tracks per Surface)	6,810	7,480
Sectors Per Track	122 to 232	63
Total Sectors	7,539,840	7,539,840

Table 10.3

- *Hard disk Partition*

    The hard disk must be *partitioned* before it can be used. *Partitions* are also sometimes called *volumes. Partitioning* a hard disk is the process of dividing its area into one or more pieces, then preparing it for use by the Operating System. Deciding how to partition a hard disk require many different considerations. Each hard disk can contain up to four different "true" partitions, which are called *primary partitions*. The limitation of four is one that is imposed on the system by the way that the *Master Boot Record (MBR)* is structured. The others are *logical partitions* that are stored in the disk's *extended DOS partition*. The extended partition counts as one of the four partitions on the disk. The extended partition is initially empty, the space can be used by adding *logical partitions* (sometimes also called *logical DOS drives* or *logical volumes*). Internally, the logical drives are stored in a linked structure. The extended partition's information is contained in the master partition table (since the extended partition is one of the four partitions stored in the master boot record). It contains a link to an *extended partition table* that describes the first logical partition for the disk. That table contains information about that first logical partition, and a link to the *next* extended partition table which describes the second logical partition on the disk, and so on. The extended partition tables are linked in a chain starting from the master partition table. In terms of how the disk is used, there are only two main differences between a primary and a logical partition or volume. The first is that a primary partition can be set as bootable (active) while a logical cannot. The second is that DOS assigns drive letters (C:, D: etc.) differently to primary and logical volumes.

- *Hard disk format and capacity*

    Most PC users are familiar with the concept that all storage media must be formatted before it can be used. There is usually some confusion, however, regarding exactly what formatting means and what it does. This is exacerbated by the fact that modern hard disks are not formatted in the same way that older ones were

    o *Two Formatting Steps:* Formatting a hard disk is actually two-step process, which is what leads to the confusion in this area. The first step, *low-level formatting* is "true" formatting. This is the step that actually creates the physical structures on the hard disk. The second step, *high-level formatting*, is an operating-system-level command that defines the logical structures on the disk such as the DOS "*format*" command on a hard disk. Partitioning is done between the two formatting steps.

    o *Low-Level Formatting: Low-level formatting* is the process of outlining the positions of the tracks and sectors on the hard disk and writing the control structures that define where the tracks and sectors are. This is often called a *"true"* formatting operation, because it really creates the physical format that defines where the data is stored on the disk. Performing a *low-level format (LLF)* on a disk permanently and completely erases it and is the closest thing to *"starting fresh"* with a new hard disk. Unfortunately, if done incorrectly, it can cause the drive to become unreliable or even inoperable. Modern hard disks are much more precisely designed and built, and much more complicated than older disks. Older hard disks had the same number of sectors per track, and did not use dedicated controllers. It was necessary for the controller to do the low-level format. Newer disks use many complex internal structures, including zoned bit recording to put more sectors on the outer tracks than the inner ones, and embedded servo data to control the head actuator. Due to this complexity, all modern hard disks are low-level formatted at the factory. Older drives needed to be re-low-level-formatted occasionally because of the thermal expansion problems associated with using stepper motor actuators. Over time the tracks would move relative to where the heads expected them to be, and errors would result. These could be corrected by doing a low-level format, rewriting the tracks in the new positions that the stepper motor moved the heads to. This is totally unnecessary with modern voice-coil-actuated hard disks. ***Warning:*** Never attempt to do a low-level format on an *IDE/ATA* or *SCSI* hard disk without forward instruction by the manufacturer during a technical support session, and if so, only using software tools that they authorize. Do not try to use BIOS-based low-level formatting tools on these newer drives.

o ***High-Level Formatting:*** After low-level formatting is completed, a hard disk has tracks and sectors, but nothing written on them. *High-level formatting* is the process of writing the file system structures on the disk that let the disk be used for storing programs and data. Using DOS, for example, the DOS FORMAT command performs this work, writing such structures as the *Master Boot Record (MBR)* and *File Allocation Tables (FAT)* to the disk. High-level formatting is done after the hard disk has been partitioned, even if only one partition is to be used. The distinction between high-level formatting and low-level formatting is important. It is not necessary to low-level format a disk to erase it: a high-level format can do that by wiping out the control structures and writing new ones, the old information is lost and the disk appears as new (typically, most of the old data is still on the disk, but the access paths to it have been wiped out). Also, different operating systems (*DOS, MAC, Windows 98, NT, UNIX*) use different high-level format programs, because they use different file systems. However, the low-level format, which is the real place where tracks and sectors are recorded, is the same.

• ***Master Boot Record (MBR)***

Whenever a PC is turned ON, the CPU has to read instruction or information (in most cases, from DRAM) then begin processing it. However, the system memory (DRAM) is empty, and the processor does not have anything to execute, or even know where it is. To ensure that the PC can always boot regardless of which BIOS is in the machine, chips and BIOS manufacturers arrange it so that the *Processor*, once turned ON, always starts executing at the same place, FFFF0H. In a similar manner, every hard disk must have a consistent "starting point" where key information is stored about the disk, such as how many partitions it has, what sort of partitions they are, etc. There also needs to be somewhere that the BIOS can load the initial boot program that starts the process of loading the operating system. The place where this information is stored in a hard disk is called the *Master Boot Record* (MBR). It is also sometimes called the *Master Boot Sector* or the *Boot Sector*. The *MBR* is always located at cylinder 0, head 0, and sector 1, the first sector on the disk. This is the consistent "starting point" that the disk always uses. When the BIOS boots up the PC, it will look in *MBR* for instructions and information on how to boot the disk and load the operating system. The *MBR* contains the following structures:

- o **Master Partition Table:** This small table contains the descriptions of the partitions that are contained on the hard disk. There is only room in the master partition table for the information describing four partitions. Therefore, a hard disk can have only four true partitions, also called *primary partitions*. Any additional partitions are logical partitions that are linked to one of the primary partitions.

- o **Master Boot Code:** The master boot record contains the small initial boot program that the BIOS loads and executes to start the boot process. This program eventually transfers control to the boot program stored on whichever partition is used for booting the PC.

    The *Master Boot Record* area is where critical and important system information is stored. If that area ever becomes damaged or corrupted then often will result in serious data loss.

- • *Volume Boot Sectors*

    Each DOS partition (also called a DOS volume) has its own *volume boot sector*. This is distinct from the *master* boot sector (or record) that controls the entire disk, but is similar in concept. Each volume boot sector contains the following:

- o **Disk Parameter Block:** Also called the *Media Parameter Block*, this is a data table that contains specific information about the volume, such as its specifications (size, number of sectors it contains, etc.), label name, etc.

- o **Volume Boot Code:** This is code that is specific to the operating system that is using this volume and is used to start the load of the operating system. This code is called by the *Master Boot Code* that is stored in the *Master Boot Record*, but only for the primary partition that is set as *active*. For other partitions, this code sits unused. The volume boot sector is created when a high-level format of a hard disk partition is performed. The boot sector's code is executed directly when the disk is booted, making it a favorite target for virus writers.

- o **Defect Mapping and Spare Sectoring:** Despite the precision manufacturing processes used to create hard disks, it is virtually impossible to create a disk with several million sectors and not have some errors show up. Imperfections in the media coating on the platter or other problems can make a sector inoperable. This usually shows up as errors attempting to read the sector, but there are other error types as well. Modern disks use *Error Checking and Correcting (ECC)* to help identify when errors occur and in some cases correct them, however, there will still be physical flaws that prevent parts of a disk from being used. Usually there are individual sectors that do not work and these are appropriately enough called *bad sectors*. When the disk drive is manufactured, it is thoroughly tested for any areas that might have errors. All the sectors that have problems or thought to be unreliable are recorded in a special table. This is called *defect mapping*. On older hard disks, these were usually recorded on the top cover of the hard disk. This information was necessary because low-level formatting was often done by the end-user and this information was used to tell the controller which areas of the disk to avoid when formatting the disk. In addition to marking them in a physical table on the outside of the disk, each defect is marked inside the drive as well, to tell any high-level format program not to try to use that part of the disk. These markings are what cause the "bad sectors" to show up when examining the disk. On modern hard disks, a small number of sectors are reserved as substitutes for any bad sectors discovered in the main data storage area. During testing, any bad sectors that are found on the disk are programmed into the controller. When the controller receives a *Read* or a *Write* for one of these sectors, it uses its designated substitute instead, taken from the pool of extra reserves. This is called *spare sectoring*. In fact, some drives have entire spare tracks available, if they are needed. This is all done completely transparent to the user, and the net effect is that all of the drives of a given model have the exact same capacity and there are no visible errors.

o ***Error Checking and Correcting or Error Correcting Code (ECC):*** *ECC* allows data that is being read or transmitted to be checked for errors and, when necessary, correct it. *ECC* is the basis for modern hard disk error detection and correction. It differs from parity checking in that errors are not only detected but also corrected. *ECC* is increasingly being designed into data storage and transmission hardware as data rates (and therefore error rates) increase. *ECC* works as follows:

1. When a unit of data (or "word") is stored in RAM or peripheral storage, a code that describes the bit sequence in the word is calculated and stored along with the unit of data. For each 64-bit word, an additional 7 bits are needed to store this code.

2. When the unit of data is requested for reading, a code for the stored and the "about-to-be-read" word is again calculated using the original algorithm. The newly generated code is compared with the code generated when the word was stored.

3. If the code matches, the data is free of errors and is sent.

4. If the code does not match, the missing or erroneous bits are determined through the code comparison and the bit or bits are supplied or corrected.

5. No attempt is made to correct the data that is still in storage. Eventually, it will be overlaid by new data and, assuming the errors were transient, the incorrect bits will "go away".

6. Any error that re-occurs at the same place in storage after the system has been turned OFF and ON again indicate a permanent hardware error and a message is sent to a log or to a system administrator indicating the location with the recurrent errors.

There are several different types of error correcting codes that can be used, but the type commonly used on PCs is the *Reed-Solomon* algorithm. This technique is also used for error detection and correction on CD-ROM media, and is also used on some systems to detect and correct errors in the system memory. At the 64-bit word level, parity-checking and ECC require the same number of extra bits. In general, ECC increases the reliability of any computing or telecommunications system (or part of a system) without additional cost. *Reed-Solomon* codes are commonly implemented; they are able to detect and restore "erased" bits as well as incorrect bits. The basis of all error detection and correction is redundant information and special software to use it. Each sector of data on the hard disk contains 512 bytes, or 4,096 bits, of user data. In addition to these bits, an additional number are devoted for use by *ECC*. The number of bits used per sector is a design decision: the more bits used, the more robust the error detection and correction, but the fewer sectors that can be fit on the track (since less of the disk's space is available for user data). Many hard disks now use over 200 bits of *ECC* code per sector. When a sector is written to the hard disk, *ECC* codes are generated and stored in their reserved bits. When the sector is read back, the user data read, combined with the *ECC* bits, telling the controller if any errors occurred during the read. Not every error can be detected and corrected, but the vast majority of the most common ones can. Sophisticated drive firmware uses the *ECC* as part of its overall error management protocols.

o **Unformatted and Formatted Capacity:** A small amount of the space on a hard disk is taken up by the formatting information that marks the start and end of sectors, and other control structures. For this reason, a hard disk true size depends on the formatted and the unformatted capacity but the only area that matters is the formatted capacity.

o **Binary vs. Decimal Capacity Measurements:** One of the most confusing problems regarding capacity measurements is the fact that the computing world has two different definitions for most of its measurement terms. Capacity measurements (*Chapter 2*) are usually expressed in Kilobytes (thousands of bytes), in Megabytes (millions of bytes), Gigabytes (billions of bytes) or terabytes (trillion of bytes). Due to a mathematical coincidence, there are two similar but different ways to express a Megabyte or a Gigabyte. Computers are digital and store data using binary numbers, or powers of two. 2^{10} (two to the tenth power) is 1,024 which is approximately 10^3 or 1,000. For this reason, 2^{10} is called a Kilobyte. Similarly, 2^{20} is 1,048,576 (a Megabyte, as is 10^6 or 1,000,000), and 2^{30} is 1,073,741,824 (a Gigabyte, as is 1,000,000,000). The numbers expressed as powers of two are called *binary* Kilobytes, Megabytes, and Gigabytes, while the conventional powers of ten are called *decimal* Kilobytes, Megabytes and Gigabytes. In many areas of the PC, only binary measures are used. For example, 64 MB of system RAM always means 64 times 1,048,576 bytes of RAM or 67,108,864. In other areas, only decimal measures are found, a 28.8 KBaud Fax/Modem works at 28,800 bits per second, not 29,491 (28.8 times 1,024).

$$\frac{1,024 - 1,000}{1,000} = 0.024 \text{ or } 2.4\%$$

$$\frac{1,048,576 - 1,000,000}{1,000,000} = 0.049 \text{ or } 4.9\%$$

$$\frac{1,073,741,824 - 1,000,000,000}{1,000,000,000} = 0.074 \text{ or } 7.4\%$$

Figure 10.1

With disks, however, some companies and software packages use *binary* Megabytes and Gigabytes, others use *decimal*. Notice that the difference in size between *binary* and *decimal* measures gets larger as the numbers get larger. There is only a 2.4% difference between a decimal and a binary Kilobyte, but this increases to a 4.9% difference for Megabytes, and around 7.4% for Gigabytes (*Figure 10.1*), which is actually fairly significant. One of the biggest problems is that hard disk manufacturers almost always state capacities in *decimal* Megabytes and Gigabytes, while most software uses *binary* numbers. This is also much of the source of confusion surrounding 2.1 GB hard disks and the 2 GB DOS limit on partition size. Since DOS uses *binary* Gigabytes, and 2.1 GB hard disks are expressed in decimal terms, a 2.1 GB hard disk can in fact normally be entirely placed within a single DOS partition. 2.1 *decimal* Gigabytes is actually 1.96 binary Gigabytes. Another example is the BIOS limit on regular IDE/ATA hard disks, which is either 504 MB or 528 MB, depending on which type. Another thing to be careful of is converting between binary gigabytes and binary megabytes. *Decimal* gigabytes and megabytes differ by a factor of 1,000 while the binary measures differ by 1,024. So this same 2.1 GB hard disk is 2,100 MB in decimal terms. But its 1.96 binary gigabytes are equal to 2,003 binary megabytes (1.96 times 1,024). Under the *IEEE* proposal, a "Megabyte" or "1 MB" is 10^6 bytes and "Mebibyte" or "1 MiB" is 2^{20} bytes.

- ***BIOS Settings - IDE Device Setup / Autodetection***

Most *BIOS* have an entry in the Standard Setup menu for each of the four *IDE/ATA* devices supported in a modern system (primary master, primary slave, secondary master, and secondary slave). For each one of the devices, a value can be entered for each setting in this section (*type, size, cylinders*, etc.). It should be noted that all modern hard disks use special technologies that makes simple geometry figures like "cylinders, heads, sectors" inapplicable. For example, almost all modern drives use a variable number of sectors, and are set up using an "approximate" figure in the system BIOS. Virtually all BIOS come with IDE device autodetection in two forms:

- ***Dynamic IDE Autodetection:*** This is the fully automatic mode. The IDE devices (primary master, primary slave, etc.) are set on "Auto" and the BIOS will automatically re-detect and set the correct options for the drive each time the PC is turned ON. The BIOS will usually display on the screen what device it finds each time it autodetects. For most people, this is the best way to go; it ensures that the *BIOS* always has the correct information about system's hardware, and it removes any possibility of installing a new drive but forgetting to set up the *CMOS* properly, or of changing a parameter by mistake in the setup program.

- ***Manual IDE Autodetection:*** This type of *autodetection* is run from the *BIOS* setup program. *Autodetection* is selected, the *BIOS* will scan the *IDE* channels, and set the *IDE* parameters based on the devices it finds. When the *BIOS'* settings are saved in *CMOS* memory, they are recorded permanently. The disadvantage of this is that if a device is changed, the *BIOS* must re-*autodetect* it (unlike the dynamic *autodetection* scheme, which does a fresh *autodetection* at startup of the PC). Virtually every *BIOS* created in the 1990's offers manual *autodetection*. When dynamic autodetection is used, the BIOS will normally "lock" the individual device settings that are being automatically set by the BIOS at boot time. Most systems that provide manual autodetection will not lock the individual settings; they autodetect, set the settings, and then allow the user the option to change them. Using autodetection for IDE/ATA devices is *strongly* recommended. It is the best way to reduce the chances of disk errors due to incorrect BIOS settings. It also provides immediate feedback of problems; if a drive cannot be autodetected from the BIOS, the user knows that there is a problem even before the PC startup.

Carlos E. Hattab

Warning:

1. If a *BIOS* contains a *"hard disk utility"* or *"low-level format"* type program, **do not use it on IDE/ATA drives**. These utilities are intended for older *MFM* and *RLL* type devices. Modern *IDE* drives do not need low-level formatting, interleave factor settings, or media analysis under all but the very most unusual circumstances, and when they do need it, they need special utilities specially designed for the type of drive it is being used.

2. Changing the mode on a hard disk, for example from Large to *LBA* or vice-versa, can change the translation method that the BIOS uses for the drive. This can also happen if the drive is moved from a computer that does not use *LBA* to one that does use it. If the translation mode changes, there is a risk of losing all the data on the drive. It is recommended that no change translation methods be made unless there is a specific reason to do so, and that the data is backed up before changing these types of settings in the *BIOS*.

o **Block Mode:** If *"enabled"*, this mode allows the system to perform accesses to the hard disk in *block mode*. More than one sector can be transferred on each interrupt; new drives allow the computer to transfer as many as 16 or 32 sectors at a time. This greatly improves performance when the system is using multitasking Operating Systems such as *Windows 95* or *Windows NT*, since the processor is *"distracted"* from its other work much less often. Normally this setting is enabled if the hard disk is compatible. Disable this mode if the system experiences lockups or problems with the hard disk or other peripherals.

o **Programmed Input/Output or PIO Mode:** This is the *PIO* mode setting for the *IDE* device. *IDE/ATA* uses one of two different ways to transfer information into and out of memory: *Programmed I/O (PIO)* or *Direct Memory Access (DMA)*. There are 5 different *PIO* modes, from 0 to 4, with 4 being the fastest (*Table 8.1*), new drives support the faster modes. For maximum PC performance, the highest mode of the drive is selected. If there are hard disk problems with the PC, it is recommended to select the next slower mode until PC is stabilized from any hard disk problems. As the *PIO* mode is slowed, it will impact performance.

o **32-Bit Transfer Mode:** Enabling this setting allows for 32-bit data transfers between the processor and the *PCI* Bus. Actual transfers to the disk are always done 16-bit at a time, but enabling this option will cause a small performance improvement on the transfer from the bus to the processor. If the hard disk supports this mode, enable it for a performance increase. Disable it, if there is any sign of PC difficulties.

• **File System Performance**

There are several factors related to how the disk is logically structured and the file system set up and maintained that can have a tangible effect on performance. These are basically independent of the hard disk and will have similar impacts on any hard disk:

o **File Structure and Cluster Size:** When using *DOS* or a *Windows* variant that uses *DOS*-based file structures, the choice of cluster size has an impact on the system's performance. In other words, larger clusters waste more space due to slack but generally provide for slightly better performance because there will be less fragmentation and more of the file will be in consecutive blocks. There is also less overhead to be maintained because the file allocation table is smaller. Similarly, a file system using *FAT32* will have different performance characteristics than one using *FAT16*.

90

o **Fragmentation:** A fragmented file system leads to performance degradation. Instead of a file being in one contiguous chunk on the disk, it is broken into many pieces, which can be located anywhere on the disk. This means forcing additional seeks when reading a large file, and seeks are very time-consuming. Defragmenting a very fragmented hard disk will often result in overall performance improvements.

o **Partitioning:** The way that a disk is partitioned can affect its performance. This is related to the cluster size issue, because smaller partitions generally mean larger clusters and vice versa. This is also affected by zoned bit recording. If a drive is partitioned into three Logical Partitions, the first one will have the best performance because it is using cylinders at the outer edge of the disk.

CHAPTER XI
BACKUP DEVICES

- ***Backup media***

"Backup" is the safety net of a total PC data loss. In the world of information, nothing is worse than losing the information data stored in a computer. Virtually every component of a PC system can fail any time and that component can be replaced easy and quick. With backup must come the confidence to retrieve the data back into the PC quickly and easily. Backup means being able to restore the computer data in the aftermath of a PC crash. A *"PC crash"* is when access to the primary hard disk of the PC is forbidden by a hardware failure, a software corruption or a virus attack. The higher the value of the data is, the more comprehensive the backup needs to be.

Figure 11.0

The Personal Computer's data backups are exactly as safe as the physical media that contains them. If the PC's data was copied onto a tape backup (*Figure 11.0*) then the tape was left lying on top of the PC box or near a magnet (*Figure 11.1*), the backup would be partially or fully lost within minutes. Once completed, backup media should be stored in a safe place, away from the PC. Tape backups should be given the same protection against risks such as theft, disaster or sabotage. In fact, the word safe is appropriate, since a locked safe is the best storage container for backup media. The use of a safe allows the media to be secured from prying eyes and fingers while remaining in the same general vicinity as the PC. A fireproof safe is an even better idea. Depending on the type of backup media used, the storage environment must be appropriate. For magnetic media such as floppy disks and tape backup, it is recommended that the storage area offer protection from the hazards that threaten them, such as temperature, moisture, dirt, water, direct sun and magnetic fields. Off-site storage of one or more of the backup media sets in the media rotation system is encouraged to allow for safeguarding against total disaster such as fire, tornado, earthquake and hurricane.

Figure 11.1

- *What To Back Up?*

Ensuring that the system's backup performed properly is an upfront protection from future frustration and nightmares. It must be determined in the beginning of the backup *what files to back up* and *how often to back them up*. Some files will need to be backed up more often than others. This decision will help facilitate the process of full and partial backups. While virtually all regular files should be backed up, in most cases the following kinds of files should be excluded from routine backups:

  o *Swap File:* A *Swap File* is a file that is used by the Operating System for virtual memory. When the system needs more memory than actually exists in the PC, it creates a virtual memory space and applications "share" the real memory by swapping pieces of memory to the hard disk. This file can be quite large depending on the system. Since it does not contain any real data, but rather is a placeholder for information in memory while the PC is running, there is no point in backing it up.

  o *Compressed Volume Files*: If volume disk compression is used, the compressed volume that is mounted, as a hard drive is stored on the host disk as a single file called a *Compressed Volume File* or *CVF*. All the files on the compressed disk are in this file. If compressed volume is used, the files should be backed up individually from the compressed volume. The *CVF* file on the host disk should not be backed up, since it contains the same information.

Some back up software will automatically deselect the file types above, unless it is requested by the user. Many types of backup software will also let the user select classes of files, by file type, that can be excluded for any reason.

- *Backup methods*

There are several different methods that can be selected for backup. Backups can offer different degrees of data storage security. Each method has a unique process; it depends on how the system is used and how often the data files are changed. These techniques are:

  o *Full Backup:* A full backup is the selection of all the files on the hard disk for backup. Only special files that should not be backed up at all are left out. This is the simplest type of backup, and yields the most complete backup image, but it takes the most time and media space to do.

  o *Selective Backup:* In a selective (or partial) backup, specific files and directories are selected to back up. This type of backup allows the user the control over what is backed up, at the expense of leaving part of the hard disk unprotected. Selective backups make sense when some files are changing much more rapidly than others, or when backup space is limited, although in many cases doing an incremental backup is better and easier.

  o *Incremental Backup:* An incremental backup is where only the files that have changed since the last backup are selected. If frequent backups are performed, this may back up the same files over and over, even ones that do not change over time. Instead, consider a mix of full and incremental backups. This gives the completion time and the tape space saving advantages of a selective backup while also ensuring that all changed files are covered. Incremental backups are supported by most backup software. Using the archive bit that exists for each file and directory, the hard disk is backed up. The backup software checks the bit to determine what files have been changed since the last backup, selects them for backup, and then clears the bit for all the files it backs up. If any files are changed, the software sets the bit again for the next incremental backup.

The type of files that must be backed up depends on what is important to the user, in terms of time, media cost, and also ease of restoration. Restoring a system that uses full backup requires less time than the other two methods. Incremental backups can require more steps, as first the full backup has to be restored and then the incremental backups, one after the other. Again, depending on how often the data is changing and how critical it is, the recommendation is to do a weekly full backup and a daily incremental backup.

- ***Backup programs and data***

Most files on a typical PC can be broken down, loosely, into being either programs or data. Their format differs in many ways but not as how they should be considered for backup. In general, data files should always be backed up. These are files that are hard to replace. Every backup should include all the data, either via a full backup of the entire hard disk, a selective backup that includes the directories where the data is, or an incremental backup. Program files are different than data files for two main reasons:

1. *Programs are static*; once installed they may not change (with a few exceptions of installation update).
2. *Programs are re-creatable*; if a program is corrupted or wiped out, it can be re-installed which is faster than restoring the original data from the original floppy or *CD-ROM* disk.

The combination of these characteristics suggests that backing up programs is less important than backing up data. A *Rule of Thumb* is that programs do not need to be backed up as often as data does. PC setups, updates and configurations, however, do take a considerable amount of time. Tweaking all the operating system settings, installing all the software, and modifying parameters to get everything working at high performance can be time consuming. Re-installing *Windows 95* and all of the applications, for example, may be fun to do the first few times but after that it can become frustrating. Also, remembering the special setups, the changes from the standard defaults, and the update downloads can be a problem. For these reasons, installed programs <u>should definitely not be ignored</u> when looking at backup. It is recommended that with the PC, "<u>complete</u>" media software must be available for disaster recovery. A *full backup* of the hard disk is still the best choice of backup, for the reason that it is the safest and easiest method to get the PC back to "before the catastrophic data loss". A combination of full and incremental backups, which will preserve most if not all the PC's hard disk, is the next choice.

Disaster Recovery

"*Disaster recovery*" refers to the process of restoring a system after a total loss of hard disk data. There are many different definitions of what a "disaster" is. This refers to any situation where there has to be a complete recreation of a system after a hard disk failure which means formatting the hard disk, re-installing all devices and applications software and then restoring all data files. Depending on what backup media (hardware and software) was available and how the initial system files were copied to the media, the recovery can be very simple or a great deal of work. There are several methods of recovering a backup of a system, among them:

- ***Incremental Backups Recovery:*** Incremental backups contain only the files that have been changed since the time that the last full backup was performed. This means that they cannot be used by themselves to perform a full recovery of a system in the event of disaster. To perform a proper recovery of incremental backup, the last full backup must be performed and all the files restored first. Then restore the first incremental made since that full backup, and then the next one, and so on, until the most recently made incremental backup is restored. If a full backup was performed once per week and then incremental every other day, the restoration, in the worst case, will have to be from seven different media sets.

- **Single-Step Recovery:** If the backup used was "single-step restore" or "one-step recovery" that included the creation of a boot disk containing the restore software that came with the package, it is the best and easiest method to recover the system. The *Single-step recovery* is the simplest way to get the PC back after a hard disk crash.

- **Manual DOS Recovery:** If the PC is in a *DOS* environment, and does not have a single-step recovery capability, a manual restore of the system can be accomplished without too much trouble. There can be a few exceptions however. The first thing that needs to done in performing a restore manually is to set up again the DOS environment; i.e., start up the PC system with a floppy boot disk, reformat the hard disk and install / set up the basic operating system files. Once the hard disk is bootable, reinstall the backup software (which includes the recovery utility) and run it to restore the PC applications and files. Follow the recovery directions that come with the backup software to make sure that everything is done right. The only possible conflict areas may be if the backup software itself is on the backup media and it is restored over top of the copy that has just been installed to do the restore.

- **Windows 3.x Recovery:** When restoring Windows 3.x, follow the *Manual DOS Recovery* steps, re-install Windows 3.x. If the permanent swap file was restored, then an error message that the "*swap file is corrupt*" may appear when Windows is run for the first time. This is nothing to be worried about as Windows will recreate the swap file and nothing will be lost as a result of this.

- **Manual Windows 95 Recovery**: Recovering from disaster under Windows 95 can be a very troublesome affair. In fact, it is the problems that users have with recovery in this multitasking environment that has prompted many to create single-step recovery features in backup software. There are two basic problems with manual recovery under Windows 95. If the software does not have this feature though, it can still avoid many of the nasty restore problems by using either of these techniques:

   o The least difficult technique is to reinstall *Windows 95*, then reinstall the backup software and do the full restore. Reinstalling *DOS* takes a few minutes, while reinstalling *Windows 95* can take an hour. This is unavoidable without using single-step recovery backup software.

   o The more trying problem is that to do a hard disk restore under *Windows 95*, there is a multitasking environment and there might be software conflicts. In particular, when files are being restored in the *Windows 95* directory of the PC, which includes the *Windows 95* Registry and many other files that are in active use while *Windows 95* is running, this will cause "*locked*" files representing what they are running now, that cannot be overwritten by what they are restoring. The result is often a corrupt Registry and an improper or aborted system recovery. Fortunately, there is a way to get around this that it can be tried, if the system has at least two disk partitions.

- **Backup Media and Devices**

There are many different methods that can be used to back up the data of the computer's hard disk. The primary difference between these methods is the device and medium that is used to store the backup. Different media (*Figure 11.2*) have different characteristics, such as capacity, speed, ease-of-use, universality, etc…

Figure 11.2

- o **Media Size Matching**: An important factor to consider when looking at backup alternatives is *matching the size* of the backup medium to the amount of data needed to backup. As hard disks continue to increase greatly in size, it is becoming more difficult to find backup solutions that can handle the entire contents of a PC using a reasonable amount of media. It is essential that the size of the backup medium be matched to the size of the data being backed up. It is tempting to ignore this issue as unimportant, but it is very clear: the more disks or tapes it takes to perform a backup, the less likely it is that they will be done on a regular basis. The best solution is a backup where the entire contents of the PC's hard disk can be stored on a single backup tape or disk. This is the ideal *unattended backup:* "start the backup, and leave the PC alone until the backup is done". If the contents of the hard disk would not fit onto a single cartridge, the backup software requests an intervention by the user at some point to change the media. This request may change the backup from a simple to a complicated one because the cartridges now have to be numbered and stored together. If any cartridge is lost so is the backup.

  - **Floppy Disks:** Floppy disks are not suitable as backup devices for a PC because they are slow, relatively unreliable and have limited capacity. Their only possible use for backup is for archiving small files. Floppy disks are importantly used as a startup system disk or a vehicle for storing critical data files about the PC system. These emergency boot disks are best stored on floppies so they can be used in the event of a hard disk problem.

  - **Tape Drives** (*Figure 11.3*): Once much maligned as too slow, too expensive and too unreliable, tape is making a comeback as a storage device in the PC world. Tape is in fact, in many cases, the best backup medium for the average PC user if used properly. The advantages of tape are:

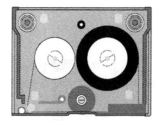

Figure 11.3

- **Capacity**: Hard drive capacity keeps getting larger and larger. Tape is the only economical backup medium that is of a similar size and growing in capacity to match them.

- **Cost**: Tape is inexpensive, both for the drives and on a per-Gigabyte basis for the media. A typical PC user can purchase a drive and enough media to do proper, reliable backups of several gigabytes of data, for a reasonable cost.

- **Reliability:** Tape is a reasonably reliable backup medium, provided that the drive is maintained properly and the media are treated with care.

- **Simplicity and Universality:** There is a lot of support for tape drives today as well as a number of software packages that will support a wide variety of devices. They are usually relatively simple to set up.

   Tape is still far from perfect. While reliability is good, it is not as good on some of the cheaper units as it is on more expensive tape drives. Performance is also less than ideal in many cases, especially when random access is needed for specific files on the tape. Fortunately, this needs to be done mostly when restoring data (backups are large, sequential writes and quite suited to the use of tape drives). New and high-end tape drives have very good performance as well. Typical access time for magnetic tape drive is between 20 and 500 seconds.

   - **Removable Storage Drives:** These have existed in various forms for many years but it is only recently that they have been getting mass market attention. Their popularity rests in their ability to provide removable storage at a reasonable price and with good performance. Many of these drives are also very suitable to use for backups. There are many different drives, and they differ in many different ways. The drives fall into several categories:

- **Large Floppy Disk Equivalent Drives**: These are the floppy-like-drives with a capacity of over 100 MB such as the Iomega Zip disk and the Maxell Super Disk (*Figure 11.4*). These devices are suitable for backup only as a partial hard disk, or if the user has the diligence and patience to do attended backups or large numbers of partial backups. As hard disks increase in size to 14 GB and beyond, trying to do backups to a device that is only a little more than 100 MB becomes impractical, and quite expensive. The reliability of these devices is quite good, although they are proprietary and not very universal. Their performance is general poor to average.

Figure 11.4

- ***Removable Hard Disk Equivalent Drives or Cartridges***: This category includes devices such as Iomega's Jaz drive and SyQuest's SyJet. Removable cartridges (*Figure 11.5*) are hard disks encased in a metal or plastic cartridge; that can be removed like a floppy disk. Removable cartridges are very fast, though usually not as fast as fixed hard disks. These are somewhat suitable for use as backup devices due to their larger capacity, but even with capacity between 500 MB to 2 GB, they are becoming inadequate for unattended backups. These drives have generally much higher performance than the smaller drives and cost more. Reliability is usually good, and the drives are still proprietary.

Figure 11.5

- ***CD-Recordable or CD-R***: These are *Write-Once Read-Many or WORM* drives with a capacity of about 650 MB (*Figure 11.6*). Despite the fact that the disks are not reusable, they are actually used for backup. While this is a very expensive way to do backups; it does give the user the advantage of being able to refer back to historical snapshots of the system's hard disk for a long time. The cost will discourage most people from doing backups often enough. The capacity is on the small side at 650 MB. One great advantage is that the backups are readable by any CD-ROM drive. CD-R is not recommended for routine backups, because the cost of this media over time is excessive.

Figure 11.6

- **CD-Rewriteable or CD-RW:** This drive is really in the same category as the CD-R. *CD-RW* is a *general-purpose* medium (*Figure 11.7*), because of its flexibility. Its media is reusable and it can also write a CD-R that can be read on most CD-ROM drives. It has the capacity of about 650 MB, and the CD-RW disks are essentially proprietary since only newer CD-ROM drives will read them. As a strictly backup medium, CD-RW is usable for backup but is not recommended because of CD-ROM drives and the cost of media over time.

Figure 11.7

- **Removable Hard Disks:** An interesting backup method that is not widely used. A *"removable hard disk"* special kit can be purchased and installed into the PC (includes an external hard disk bay). A special adapter (sometimes called a *carrier*) is attached to the computer internal IDE hard disks. This allows the PC to insert and remove these internal disks through an external drive bay, converting the PC's IDE hard disk into a removable drive. Complete hard disk duplication can be done between the internal PC hard disk and the external removable hard disk. A removable hard disk has the following advantages:

  a. Very high performance.

  b. Random-access capability.

  c. Standard interfaces and exchangeability.

99

    d. Excellent reliability.

The disadvantages:

    a. Compared to tapes, removable hard disks are very expensive.

    b. Hard disks are also fragile; drop them and they may be damaged.

    c. Disks can only be removed when the power is OFF.

- ***Hard Disk Duplication:*** Another backup for PC with more than one hard disk use is to set up the secondary hard disk to regularly duplicate (file or disk copy tools) the primary hard disk. The advantages:

    a. Simple.

    b. Can be automated.

    c. Performance of the disk-to-disk copy will be very high.

    d. Cost is reasonable.

The disadvantages:

    a. No protection against risks to data from hazards such as theft, fire, sabotage, many types of viruses, and even some types of hardware failure.

    b. Single backup is made at a time, which makes the whole system very vulnerable. If a problem occurs and wipes out some or all the data, it is a crisis.

    c. Temptation, at some point, to use the second drive for more data and discontinue the backup procedure when the first disk gets filled up.

       Hard drive duplication as a *standalone* backup procedure is not recommended. It can be useful when supplemented by a removable backup system. A different PC backup with more than one hard disk can set up the primary hard disk for system, applications and programs files. The secondary hard disk can be for data. The backup is now focused on the secondary hard disk only.

- ***Network Backup:*** For PCs on a network, backup over the network is a viable alternative to using removable drives. This type of backup is sometimes used in small- to medium-sized backup as a way of protecting PCs without the expensive of tape drives or removable storage. The idea is fairly simple: copy data from one PC to another over the network. Duplicating each PC's information provides a way to protect each individual PC. In a way, this type of backup is most similar to *Hard Disk Duplication* in terms of how it works. It is simple in the same way, and can be automated. It addresses some of the concerns about that method: there is not the same single point of failure in terms of virus attack or hardware failure. However, depending on the location of the PCs, theft, disaster and sabotage can still be a big problem, even if the PCs are located in different areas (offices, factories, etc.). In addition, a file-infector viruses can travel over a network and this is a real threat. An even better use of the network, becoming popular in many corporate environments, is to use a centralized removable storage backup device in conjunction with the network to back up all the PCs automatically. It is a local-area network with many PCs connected to a server that has an 8 mm tape drive. Using compression, this drive can hold the entire contents of the network on a single tape, and using network backup software and the right operating system; the network can be used to back up all the PCs every night, automatically. This is an excellent backup system that allows the desktop PCs to reap the benefits of the tape backup unit without everyone having to remember to do backups.

- ***File Archiving:*** A recommended supplemental backup method is *file archiving*. This is simply making backup copies of files that are periodically modified. In this case, the practical backup is to copy the file to another location on the hard disk once in a while. This is a limited form of *Hard Disk Duplication*, which is not a complete backup solution. In fact, this system only really protects well against accidental deletion, which is why it is called *supplemental*.

o ***Backup method comparison:***

Comparison charts are shown in *Table 11.0* and *Table 11.1*. The scale is from High (better) to Moderate (acceptable) to Low (worse).

Data Risk	Floppy Disks	Tape Drives	Removable Storage Drives	Removable Hard Disks	In-Place Hard Disk Duplication	Network Backup	File Archiving
Hardware Failure	High	High	Moderate to High	Moderate to High	Moderate	High	Low
Software Failure	High	High	High	High	Moderate	High	Low
File System Corruption	High	High	High	High	Low to Moderate	Moderate to High	Moderate
Accidental Deletion	High	High	High	High	High	High	High
Virus Infection	Moderate	High	Moderate to High	Moderate	Low	Moderate to High	Low to Moderate
Theft	High	High	High	Moderate to High	None	Low to Moderate	None
Sabotage	High	High	High	High	Very Low	Low	None
Natural Disaster	High	High	High	High	None	Low to Moderate	None

Table 11.0

Characteristic	Floppy Disks	Tape Drives	Removable Storage Drives	Removable Hard Disks	In-Place Hard Disk Duplication	Network Backup	File Archiving
Capacity	Very Low	High	Low to High	High	High	High	--
Automation	Low	Low	Low	Low	High	High	High
Initial Cost	Very High	Low to High	Low to Moderate	Moderate	Moderate	High	Very High
Media Cost	Low	Moderate to High	Low to Moderate	Moderate	Very Low	Very High	Very High
Expandability	High	High	High	Moderate to High	Low	Very High	High
Reliability	Low	Low to High	Moderate to High	Moderate to High	High	Very High	Very High
Simplicity	Moderate	Moderate to High	Moderate to High	Low	Very High	Moderate	High
Universality	Very High	Low to High	Low to Moderate	Moderate	High	High	High
Performance	Very Low	Low to Moderate	Low to High	Very High	Very High	Moderate to High	Very High

Table 11.1

- ▪ ***Characteristics:*** The characteristics of tape backup include the followings:

  - *Capacity:* Tape storage size (MB, GB).

  - *Automation:* Backup automation.

  - *Initial Cost:* Hardware and software startup cost.

  - *Media Cost:* Once startup complete, additional cost.

  - *Expandability:* Additional backup media.

  - *Reliability:* Restore from a disaster loss.

  - *Simplicity:* Backup methods.

  - *Universality:* Universal hardware / media.

  - *Performance:* Hardware / software overall performance.

- ▪ ***Data risk coverage:*** Some backup methods and devices do a better job than others of protecting against the risks to PC programs and data. *Table 11.1* shows a general summary of how various methods rate in terms of protecting the PC from the hazards that threaten valuable programs and data. *Table 11.1* is a guideline that must be evaluated by the user depending on how the PC is used. The ultimate decision on what is the best choice for backup is with the user:

- ***Interface speed***

The objective of using a hard disk is to transfer data to and from the disks. This involves two processes:

1) ***Internal Process:*** It is the actual reading or writing the media.

2) ***External Processes:*** Moving the data from the inside of the drive out to the system, or vice-versa.

o ***Data Transfer Rate***

- ▪ ***Internal Transfer Rate*** also called *Sustained Transfer Rate*, respectively. When dealing with a large transfer, the *Sustained Transfer Rate* is the limiting factor is how fast the disk's internal mechanisms can operate.

- ▪ ***External Transfer Rate*** also called *Burst Transfer Rate, Interface Transfer Rate* or *Host Transfer Rate,* is the speed of transfer over the interface between the hard disk and the rest of the PC. It is called *Burst* because it can be much higher than the sustained rate, for short transfers. The *External Transfer Rate* is the speed at which data can be exchanged between the system memory and the internal buffer or cache built into the drive. This is usually faster than the internal rate because it is a purely electronic operation, which is much faster than the mechanical operations involved in accessing the physical disk platters themselves. This is in fact a major reason why modern disks have an internal buffer. Primarily the type of interface used, and the mode that the interface operates in dictates the External Transfer Rate.

Support for a given mode has two requirements: the drive itself must support it, and the system (usually meaning the system *BIOS* and chipset) must support it as well. Support for the higher transfer modes also usually requires that the interface be over a high-speed system bus such as *VLB* or *PCI*, which most today are. The two most popular hard disk interfaces used today are *SCSI* and *IDE/ATA* (and enhancements of each). *IDE* uses two types of transfer modes: *PIO* and *DMA*. *Table 11.2* shows a comparative interface summary:

Interface	Mode	Theoretical Transfer Rate (MB/s)
IDE/ATA	Single Word DMA 0	2.1
IDE/ATA	PIO 0	3.3
IDE/ATA	Single Word DMA 1	4.2
IDE/ATA	Multiword DMA 0	4.2
Standard SCSI	--	5
IDE/ATA	PIO 1	5.2
IDE/ATA	PIO 2	8.3
IDE/ATA	Single Word DMA 2	8.3
Wide SCSI	--	10
Fast SCSI	--	10
EIDE/ATA-2	PIO 3	11.1
EIDE/ATA-2	Multiword DMA 1	13.3
EIDE/ATA-2	PIO 4	16.6
EIDE/ATA-2	Multiword DMA 2	16.6
Fast Wide SCSI	--	20
Ultra SCSI	--	20
Ultra ATA	Multiword DMA 3 (DMA-33)	33.3
Ultra Wide SCSI	--	40

Table 11.2

For high performance, it is important that the external (*burst*) transfer rate be higher than the internal (*sustained*) transfer rate of the drive. The drive should be used to its maximum potential and the interface should be upgraded (e.g., *EIDE, Ultra SCSI, etc.*).

• ***Backup scheduling and media rotation***

In order to provide maximum safety for the PC's data, it is important to plan out a backup schedule that will allow the user the most flexibility and reliability in recovering from potential disasters. This means, in almost every case, the use of multiple backup media and a backup schedule that dictates when each set of media should be used. The media set could be magnetic tapes, CD-Rs, CD-RWs and removable cartridges. The amount of media needed depends on when protection recovery is required, and what sort of retention period is needed to maintain the PC data.

o **Retention Period:** The *Retention Period* refers to the amount of history data needs to be stored in a media. The longer the data is kept in a media before reusing it, the longer the retention period. Longer periods protect the PC better against problems that are gradual and take a while to notice. They give the use more flexibility to go back and see how things looked in the past. If a single backup tape was setup and used for every backup, the retention period is extremely short. If a snapshot of the entire hard disk was stored on a CD-R disk every month for the last three years, files can be retrieved quickly. For some PC users, very long retention periods are not that important; they need only enough to enable them to deal with problems that may take some time to notice. For example, if the hard disk was infected with program or data virus, and the PC retention period is only about a week, there is a real risk of having the "virus-contaminated programs" not only on the hard disk, but also on the tape backups as well.

o **Media Rotation Tradeoffs:** The reason that there are different rotation schemes is that there are different systems, and different people who use them, and therefore different needs. The backup requirements of a home PC are very different than the server for a business. There is also a matter of media cost, which in turn depends on the type of backup being done. In general, these are the tradeoffs:

  ▪ **Retention Period:** Using more media sets gives the user a higher retention period, protecting the PC against various kinds of problems better.

  ▪ **Media Failure Protection:** The more media is used, the less likely there will have a failure with the backup device that cripples the system. It is also possible to lengthen the life of each individual backup disk or tape.

  ▪ **Cost:** In general, using more media sets costs more money. This can be an important factor in deciding on a backup unit. An Iomega "Jaz" drive is fast and has random access, but many backup cartridges are expensive. One reason that professionals use 8mm tape drives or similar for backup is that the media cost for 8mm tapes is less of that for removable drives like the "Jaz".

  ▪ **Restore Time:** Systems that rely on incremental backups take much more time to restore in the event of a disaster.

o **Minimum Security Rotation System:** The absolute minimum backup schedule is the use of two media sets, and rotating them on alternate backups. This gives at least some protection against relying on a single backup media set, and eliminates the chance of a hard disk crash during backup resulting with nothing at all. It is still not the best protection, since the retention period is very low, and there is no provision for off-site media storage. If this rotation is used, a decision needs to be made on how often to do the backup. The longer the time interval between backups, the higher the risk on the changes made since the last backup. For many people who use their systems little enough that this rotation is sufficient. For others, daily backups make more sense. If daily backups are needed, however, a more disciplined rotation is better: daily incremental and weekly full backups.

o **Light Security Rotation System:** This rotation is suitable for a PC that gets light to moderate use. It provides better protection and a longer retention period than the minimum security system, while keeping the number of media sets small and the number of backups performed low. This system uses four media sets, labeled "Mon", "Wed", "Fri 1" and "Fri 2". Starting on the first Friday, a full backup is done to "Fri 1", and then it is stored off-site. On the following Monday, an incremental backup is done to the "Mon" set, and on the following Wednesday, an incremental to the "Wed" set. When the next Friday comes around, use "Fri 2", store it off-site and bring "Fri 1" back on-site for use the next week. This rotation scheme keeps the media cost for backups to a minimum, provides off-site protection and some redundancy. It does however expose the data changed during the week to some risk, so it is recommended only for lightly-used systems.

o **Medium Security Rotation System:** This is an average media rotation system that provides good protection while requiring only a moderate number of media sets. It is suitable for a PC that gets a reasonable amount of use, and for which daily backups are a necessity. This would include a heavily used personal machine, or perhaps a smaller business PC or departmental server. Here are the particulars of this system:

- **Backup Method and Type:** Backups are performed on a tape backup unit or other removable drive with sufficient capacity to hold a full backup on one unit.

- **Media Set Groupings:** The system uses two different groups of media sets. The first group is for daily backups, the second for weekend backups. If the system is used only on weekdays, 4 daily media sets are needed, labeled "Mon", "Tue", "Wed", and "Thu". If the system is used on weekends as well, 2 more daily media sets should be added, labeled "Sat" and "Sun". Then also a number of weekly media sets should be used.

- **Weekly Full Backups:** Full backup can be performed any day of the week; suppose it is on Fridays. Label the media set "Fri 1" and store it off-site. On subsequent Fridays, backup the next media set, "Fri 2", store it off-site and return to "Fri 1". Alternate between "Fri 1" and "Fri 2". The more weekly media set that can be used, the longer the retention period will be.

- **Rotational Daily Backups:** On days other than Friday, when a daily backup is performed to the media set should be labeled with the correct day of the week. This can be either a full backup or an incremental backup. Incremental backups save time, but if a failure occurs, the restoration has to begin first with full Friday backup and then each incremental daily backups after the full backup.

- **Drive Cleaning:** The media device should be cleaned regularly for error free backup. Assuming a tape drive is used, then the drive's head should be cleaned weekly using a cleaning tape.

- **Media Storage:** It is highly recommended that all backup media sets be stored in a fire proof, locked safe, on-site. One of the weekly full backup sets should be stored off-site.

o   ***Maximum Security Rotation System:*** This rotation is most suitable for computers where large quantities of data is changing daily, the data is absolutely critical and cost of media is inconsequential compared to maximizing the chances of easy recovery in case of disaster. The system also provides for a very lengthy retention period. It requires a lot of backup media, making it only suitable for a backup system that uses inexpensive media (generally, tape). A backup recommendation rotation for such a system is:

- **Backup Method and Type:** Full Backups are performed on tape media sets.

- **Media Set Groupings:** The system alternates two different groups of media sets. The first group is for short-term rotational daily backups, the second for long term monthly and yearly backups. The number of tapes used depends on the retention period needed.

- **Rotational Daily Backups:** Daily backups media sets are rotated and last day media sets are stored off-site.

- **Monthly Backups:** On the month-end date the daily rotation is skipped for one day and long-term monthly backup media sets are used. If the month-end date is also year-end, then long-term yearly media sets are used, otherwise, the monthly media sets are selected. In both cases, the oldest permanent backup is rotated.

o   ***Backup record and maintenance***

- **Backup Logging:** A log is kept near the media drive, and on it is recorded the date of each backup and which media set was used. Each media set has the date recorded on it when it was used.

- **Drive Cleaning:** The media drive head is cleaned weekly or bi-weekly using a cleaning kit specifically designed for it. The cleaning date is logged.

- **Media Storage:** All backup media sets are stored in a fire proof, locked safe, on-site. One of the daily backup media sets is stored off-site at all times.

- **Media Replacement:** All of the daily rotation media sets are replaced once every year or two (depending on usage), to greatly reduce the chance of media deterioration.

With confidence in backup media, a computer system can provide the following advantages:

- ***Resistance to media failure:*** There is virtually no "great deal" of data being lost, due to the number of media sets used in the system. Since incremental backups are not used, there is a great deal of data redundancy on the backup media sets.

- ***Excellent Retention Period:*** There is comfort in knowing that a history of data can be retrieved from the media sets even though it might have been erased from the computer. The retrieval can go back as long as the media sets allow (a month, a year or more).

- **Rapid Restoration:** Since incremental backups are not used, a restore in the event of a problem will be much more rapid since it can be done from a media set.

- **Disaster Protection:** A safe is used for storing media. Since the media sets are rotated off-site, there is increased protection against total loss of the media sets.

- **How To Back Up**

When performing PC backups, specific techniques and considerations must be looked at as to *how to backup*. Backup timing, scheduling and media storage are some of these considerations. It is also important to note how to ensure that system backups work and how to backup with confidence.

- o **Backup Timing:** The backup timing, most often, is a personal selection (morning, evening, after midnight). It depends on how busy the PC is and also on how long will it take to perform backups. Some of the common times to backup are:

  - **Nighttime:** Unattended or when the PC is not being used by the user, is the backup choice for most systems. A full or partial backup can be pre-programmed. This is ideal for most PC users because the next morning, the backup, which can take several hours, is complete and the PC is ready to be used.

  - **Daytime:** Attended. It is a manual backup where multi-media sets need to be exchanged, a first time backup media, "on-demand" request, etc. This is a confidence and secured backup choice.

    The best *"Rule of Thumb"* would be to use the *Daytime* full backup the first time, and from then on a *Nighttime* backup.

- o **Backup Software:** When data is a valuable asset of the PC, it is important that the backup software can be trusted. A dependable software backup should have some of these features:

  - **Device Support:** Backup software varies in its ability to support backup devices. Sometimes, it is difficult to find software support for new devices. Some software developers will make software updates available for their users to provide expanded support as new drives hit the market; others will not.

  - **Operating System Support:** The software should support all the requirements and the features of the Operating System that it is running under. For example, Backup software based on *Windows 95* platform should have full support for long filenames, the *Windows 95 Registry*, and *FAT32* partitions.

  - **Backup Type Selection:** Good backup software should allow the choice between *Full, Selective* and *Incremental backups*, as well as capabilities of selecting files, subdirectories and directories based on search strings or patterns.

  - **Media Spanning:** The backup software should provide proper support for backing up to multiple pieces of media in a media set. The software should set the system to prompt the user when it is time to exchange media and when the backup is completed.

- **_Disaster Recovery:_** A very important feature, and one that is often found only on more expensive products is support for automatic disaster recovery. With this type of software, sometimes called *one-step recovery* or *single-step restore*, a floppy disk is created with a special recovery program that restores the system. Without this feature, it is often that the entire Operating System will have to be re-installed before it is restored, which could cost a lot of time and cause a lot of problems.

- **_Scheduling and Automatic Operation:_** Depending on how and when the backups are done, it can be very helpful to have the software run automatically at a preset time. Most of today's software will support this.

- **_Backup Verification:_** Every decent backup package will allow a verification mode to be enabled. When active, the software will read back from the tape every file that it backs up and compare it to the file on the hard disk, to ensure that the backup is correct. It is important to ensure that the backups are viable.

- **_Compression:_** Good backup software will give the option of enabling software compression, possibly at various levels, to save space on the backup media.

- **_Media Append and Overwrite:_** In this case, it must be able to set the software so that the user can control easily what happens when the software starts a backup of a tape that already contains a backup set. The software must always append to the tape, always overwrite it, or prompt the user each time to select it.

- **_Tape Tools:_** If using a tape backup unit, the backup software will allow formatting, rewinding, retensioning or viewing the catalog on the tape. The tape drive may come with software that perform these functions for a particular model; it is much easier if the backup software supports these tools also, however.

- **_Security:_** Better software packages will let password-protect a backup set so that the password is required to view or restore from the backup image.

- **_Backup Configuration Profiles:_** Different types of backups may be needed at different times. For example, a selection of compressed *ZIP* files on one drive to be backed up with tape compression OFF (since compression would not do anything anyway) while the regular files on another drive are backed up with compression. Some backup software will allow storage of different profiles for different types of backups to save the user from having to change things every time.

- **_General Quality Issues:_** Knowledge about the general nature of the backup software is important. Does it work well? Is it buggy? Are people having problems with it? What is the warranty? What is the upgrade policy of the manufacturer?

o **Backup Compression:** Most backup systems support some type of compression. The concept of compression is simple:

1. **Save media space:** This allows the backup of more data onto a given media set.

2. **Save backup time:** This completes the backup task in the shortest time possible.

Data compression is supported by most backup software. As a marketing attraction, many backup devices advertise their backup capacity based on the compression which will be used during backup. Data compression is recommended in many cases, however, there are some issues to consider:

- **Compressability:** Not all files will compress equally well. Files that are already in a compressed format cannot be compressed again.

- **Proprietary Formats:** Backup software is usually in a proprietary format. Backup programs use different compression algorithms. The media set written by one compression program may not be readable by a different one. Compatibility among backup software developers is not available. If compression is not used, the backup formats, however, may be reasonably universal.

- **Processing Power Requirements:** In order to write some types of backup devices, especially tape drives, it is necessary for the software to provide a steady stream of data to the device. Tape can only be written when the data is streaming at a constant speed. Compression algorithms take time to run and can sometimes interfere with this steady flow of data, causing problems such as going back-and-forth (the tape has to repeatedly back up and restart sections of the backup). If it is suspected that compression overhead is causing problems, it is recommended that the compression should be disabled.

- **Exaggerated Compression Ratios:** Be aware of the highly optimistic "estimated compression ratio" claimed by the backup device. Many media manufacturers like to claim that their compression ratio is 2:1. That means a 1.6 GB tape for example will hold 3.2 GB "with compression". That is not always true, it depends on the backup files compression ratio from the PC.

o **Disk Compression**

Disk compression techniques have been used for many years. In the 1980's, when hard disk capacity was an issue for PC users (10-100MB size), there was compression software sometimes available that would double the size of the hard disk such as "*Stacker*". Software compression allows the PC to store more information in the same amount of disk space by using special software that reorganizes the way information is stored on the disk. Hardware compression exists but is not generally used for hard disk volumes.

On most data files, disk compression takes advantage of two characteristics:

1. Most files have a large amount of redundant information, with patterns that tend to repeat. By using "placeholders" that are smaller than the pattern they represent, the size of the file can be reduced.

2. While each character in a file takes up one byte, most characters do not require the full byte to store them. Each byte can hold one of 256 different values, but if this is a text file, there will be very long sequences containing only letters, numbers, and punctuation. Compression programs use special formulas to pack information like text so that it makes full use of the 256 values that each byte can hold. The combination of these two effects results in text files often being compressed by a factor of 2 to 1 or even 3 to 1.

Compression is useful in a hard disk's slack (*Chapter 19*). For example, there are 1,000 files on a hard disk that has about 500 bytes in size but it is occupying the space of 16,384 byte clusters, which is 16 MB of disk space. In other words, the hard disk has 16 Megabytes of used space to store 500 KB or less of data. The reason is that each file must be allocated a full cluster (16,384 bytes) and only 500 of that actually has data, the remaining 15,884 bytes or 97% of that space is *slack* (*Chapter 19*). If the 1,000 files are compressed, like a "ZIP" file, not only will they probably be reduced in size greatly, but the ZIP file will have a maximum of 16,383 bytes of slack by itself, resulting in a large amount of saved disk space.

Compression of the hard disk's files can be accomplished in many different ways. There are logical mechanisms for performing the compression and decompression of data files. Some of the compression types are:

- **Utility-based file compression:** A very popular form of disk compression, used by many PCs, is a file-by-file compression using a compression utility. With this type of compression, a specific utility is used to compress one or more files into a *compressed file* (called an *archive*), and another similar utility is used to decompress the *compressed file*. The Operating System, which usually does not know anything about the compression type, views the compressed file as a typical file. In order to use the compressed file at all, it must be decompressed using a "*compress/decompress*" file tool. Usually software files that are downloaded from the Internet for example, use some type of compression.

- **Operating System file compression:** Some Operating Systems supports the compression of files within themselves. For example, Windows NT supports file compression when using the NTFS file system. This is in many ways the best type of compression, because it is both automatic and it allows full control over which types of files are compressed. Decompression, in NTFS, is done whenever any program needs a particular file.

- **Volume compression:** Distinctly different than compressing single files, it is also possible with most new Operating Systems to create entire disk volumes that are compressed. Volume compression allows the PC to save disk space without having to individually compress/decompress files. Every file that is copied to the compressed volume is automatically compressed and then each file is automatically decompressed when any software program needs it.

Carlos E. Hattab

Disk volume compression works by setting up a *virtual volume*. In essence, a software driven volume is created on the system and special drivers are used to make this volume appear to be a physical hard disk. Many devices use compression software drivers to allow them to appear to the Operating System as a hard drive. To create a compressed volume on the hard disk, the compression software, as it is creating it, may do the following:

1. Ask, the real disk partition to hold the compressed volume This sometimes is called the *host volume* or *host partition*. It will also ask whether or not to compress the existing data on that volume (if any), or instead use the current empty space on the volume to create a new compressed volume.

2. The target disk volume is prepared for compression by scanning it for logical file structure errors such as lost clusters and also for errors reading the sectors on the disk. If the disk is highly fragmented, it may need to be defragmented as well, since the compressed volume must be in a contiguous block on the disk.

3. A special file on the hard disk is created and called a *compressed volume file* or *CVF*. This file contains the compressed volume. If a compressed volume is created from empty space, the *CVF* is written directly onto the hard disk and prepared with the correct internal structures for operation. If a compressed volume is created from an existing disk with files on it, the software may not have enough free space to create the full *CVF*. It will instead create a smaller one, move some files into it from the disk being compressed, and then use the space that these files were using to increase the size of the *CVF* to hold more files. This continues until the full disk is compressed. This operation can take a very long time.

4. The *CVF* is hidden from view using special file attributes. Special drivers are installed that will make the *CVF* appear as a new logical disk volume the next time the system is rebooted. This process is sometimes called "*mounting the CVF*", in analogy to the physical act of mounting a physical disk.

CHAPTER XII
OPTICAL STORAGE DEVICES

- *Optical Devices*

Optical devices were introduced from *Laser Technology* development. Optical disks (*Figure 12.0*) are usually much faster than hard disk drives but are fairly slow compared to semiconductor *RAM*. The advantage of optical devices is their large capacity (*CD ROM* can hold 650MB).

Figure 12.0

- *Optical Disk*

An optical disk is a storage medium from which data is read and which is written on by lasers. Some Optical disks *(DVD)* can store up to 17GB (Gigabytes) of data. There are three basic types of optical disks:

  o *Compact Disc, Read Only Memory (CD-ROM):* Introduced in the 80's as a media device of the PC, the *Compact Disk (CD)* is an optical disk that is replacing the conventional 3½ floppy disk, because a single CD has the storage capacity of over 450 floppy disks.

  o *Write-Once, Read-Many (WORM):* WORM disks can be written on once, after that, the *WORM* disk behaves just like a *CD-ROM*. It can be read any number of times. A special WORM disk drive is needed to write data onto a *WORM* disk.

  o *Erasable Optical (EO):* EO disks can be read from, written to, and erased like magnetic hard and floppy disks.

These three technologies are not compatible with one another; each requires a different type of disk drive and disk. Even within one category, there are many formats, although *CD-ROMs* are relatively standardized in size. Optical disks record data by burning microscopic holes on the surface of the disk with a laser light. To read the disk, another laser beam shines on the disk and detects the holes by changes in the reflection pattern.

In audio technology, CDs also replaced the traditional *Long Play (LP)* audio discs for reliability and digital sound. The *DVD* is also a CD. The *CD-ROM*, (also known as CD), is a storage medium that uses typical CD discs to read computer data or audio files. The CD is a thin-pitted metal disc that is sandwiched between layers of plastic and lacquer. The pits do not reflect the laser light as well as the smooth surface area. A typical CD can store up to 650MB (Megabytes) of computer data or 74 minutes of audio information. Most optical disks are read only, i.e. they are already filled with data. Data can be read from the CD, but it cannot be modified, deleted or rewritten. Typical access time for *CD ROM* is between 80 and 800

113

milliseconds. Like audio *CDs*, *CD-ROMs* come with data already encoded onto them. The data is permanent and can be read any number of times, but *CD-ROMs* cannot be modified. *CD-ROM* drives play a significant role in the following essential aspects of the computer system:

- ***Software Support***

Most software today is delivered to PC users on *CD-ROMs*. At one time there were a few titles that came on *CD-ROM*, and they generally came on floppy disks as well. Today, not having a *CD-ROM* means losing out on a large segment of the PC software market. Also, some *CD-ROMs* require a drive that meets certain minimum performance requirements.

- ***Performance***

Since most software use the *CD-ROM* disks, the performance level of the drive is important. The performance of the hard drive or system components such as the processor or system memory may not be similar to that of the *CD ROM*, but it is still important, depending on what the *CD* drive is used for.

In general, the *CD-ROM* of the PC operates as follows:

1. A beam of light energy is emitted from an infrared laser diode and aimed toward a reflecting mirror. The mirror is part of the head assembly, which moves linearly along the surface of the disk.

2. The light reflects off the mirror and through a focusing lens, and shines onto a specific point on the disk.

3. A certain amount of light is reflected back from the disk. The amount of light reflected depends on which part of the disk the beam strikes: each position on the disk is encoded as a one or a zero based on the presence or absence of *"pits" (hole)* in the surface of the disk.

4. A series of collectors, mirrors and lenses accumulates and focuses the reflected light from the surface of the disk and sends it toward a photo detector.

5. The photo detector transforms the light energy into electrical energy. The strength of the signal is dependent on how much light was reflected from the disk.

Most of these components are fixed in place; only the head assembly containing the mirror and read lens moves. This makes for a relatively simplified design. *CD-ROMs* are single-sided media, and the drive therefore has only one "head" to go with this single data surface. Since the *read head* on a *CD-ROM* is optical, it avoids many of the problems associated with magnetic heads. There is no contact with the media as with floppy disks so there is no wear or dirt buildup problem. There is no intricate close-to-contact flying height as with a hard disk so there is no concern about head crashes and the like. However, since the mechanism uses light, it is important that the path used by the laser beam be unobstructed. Dirt on the media can cause problems for *CD-ROMs*, and over time dust can also accumulate on the focus lens of the read head, causing errors as well. The lens assembly that moves across the *CD-ROM* media is similar to the heads on a hard disk or floppy disk drive. The technology used to move the read head on a *CD-ROM* drive is in some ways a combination of those used for floppy disk drives and for hard disk drives. Mechanically, the head of the *CD-ROM* moves in and out on a set of rails, much as the head of a floppy disk drive does. At one end of its travel the head is positioned on the outermost edge of the disk, and on the other end it is near the hub of the CD. However, due to the dense way the information is recorded on the CD, *CD-ROM* drives cannot use the simple stepper motor positioning of a floppy disk. *CD-ROM* media actually use a tighter density of tracks than even hard disks do. The positioning of the head is controlled by an integrated microcontroller (CPU and memory chip) and servo system. This is similar to the way the actuator on a hard disk is positioned. This means that the alignment problems found on floppy drives (and older hard disks) are not generally a concern for *CD-ROM* drives, and there is some tolerance for a CD that is slightly off center. Like a floppy disk, the head actuator on a *CD-ROM* is relatively slow. The amount of time taken to move the heads from the innermost to the outermost tracks (called *Full Stroke Seek*) is about an order of magnitude higher than it is for hard disks.

1. **Spindle Motor**

Similar to all spinning-disk media, the *CD-ROM* drive includes a spindle motor that turns the media containing the data to be read. The spindle motor of a standard *CD-ROM* is very different from that of a hard disk or floppy drive in one very important way: it does not spin at a constant speed. Rather, the speed of the drive varies depending on what part of the disk (inside vs. outside) is being read. Standard hard disks and floppy disks spin the disk at a constant speed. Regardless of where the heads are, the same speed is used to turn the media. This is called *Constant Angular Velocity (CAV)* because it takes the same amount of time for a turn of the 360 degrees of the disk at all times. Since the tracks on the inside of the disk are much smaller than those on the outside of the disk, this constant speed means that when the heads are on the outside of the disk they will traverse a much longer linear path than they do when on the inside. Hence, the linear velocity is not constant. Newer hard disks take advantage of this fact by storing more information on the outer tracks of the disk than they do on the inner tracks, a process called *zoned bit recording*. They also have higher transfer rates when reading data on the outside of the disk, since more of it spins past the head in each unit of time. *CD-ROMs* take a different approach. They adjust the speed of the motor so that the linear velocity of the disk is always constant. When the head is on the outside of the disk, the motor runs slower, and when it is on the inside, it runs faster. This is done to ensure that the same amount of data always goes past the read head in a given period of time. This is called *Constant Linear Velocity (CLV)*. The reason that *CD-ROMs* work this way is based on their heritage of being derived from audio CDs. Early CD players did not have the necessary buffer memory to allow them to deal with bits arriving at different rates depending on what part of the disk they were using. Therefore, the CD standard was designed around *CLV* to ensure that the same amount of data would be read from the disk each second no matter what part of it was being accessed. *CD-ROMs* were designed to follow this methodology. The speed of the spindle motor is controlled by the microcontroller, tied to the positioning of the head actuator. The data signals coming from the disk are used to synchronize the speed of the motor and make sure that the disk is turning at the correct rate. The first *CD-ROMs* operated at the same speed as standard audio CD players: between 210 and 539 *RPM (Rotation Per Minute)* depending on the location of the heads. This results in a standard transfer rate of 150 Kilobytes/second. It was realized fairly quickly that by increasing the speed of the spindle motor, and using sufficiently powerful electronics, it would be possible to increase the transfer rate substantially. There is no advantage to reading a music CD at double the normal speed, but there definitely is for data CDs. Thus the double-speed (two times), or 2X *CD-ROM* was introduced. It followed in short order with 3X (three times), 4X (four times) and faster drives (CD drives with speed over 48X are available today).

Virtually all *CD-ROM* drives (up to 12X speed) vary the motor speed to maintain constant linear velocity. As the speed of the drives has increased, many new drives have come out to actually revert back to the *CAV* method used for hard disks. In this case, their transfer rate will vary depending on where on the disk they are working, again, just like it does for a hard disk. Some drives use a partial *CLV* or mixed *CLV/CAV* methods to enhance the speed of the disk. This is a compromise design that uses *CAV* when reading the outside of the disk, but then speeds up the spin rate of the disk while reading the inside of the disk. This is done to improve the transfer rates at the inside edge of the disk. The change back to *CAV* as the drives get faster and faster is being done due to the tremendous difficulty in changing the speed of the motor when it is going fast. The change in a disk spinning at 210 RPM to 539 RPM and back again is actually one factor contributing to the slow performance of CD-ROMs especially on random accesses. *Table 12.0* summarizes the differences between *CLV* and *CAV*:

CLV and CAV Comparison

Characteristic	Constant Linear Velocity (CLV)	Constant Angular Velocity (CAV)
Drive Speed	Variable	Fixed
Transfer Rate	Fixed	Variable
Application	Conventional CD-ROM drives	Faster and newer CD-ROM drives, hard disk drives, floppy disk drives

Table 12.0

2. Loading Mechanism

The *loading mechanism* refers to the mechanical components that are responsible for loading CDs into the CD-ROM drive. There are two different ways that CD-ROM media are normally loaded into the CD-ROM drive:

1. *Tray:* The *tray* is the most popular loading mechanism used today. With this system, a plastic tray, driven by gears, holds the CD. When the eject button is pressed the tray slides out of the drive, and the CD is placed upon it. The tray is then loaded back into the drive when the eject button is pressed a second time. Most drives will also respond to a slight "push" on the drive tray by activating the mechanism and retracting the tray.

2. *Caddy:* Many older and some new CD-ROM drives use *caddy*. This is a small carrier made of plastic. A hinge on one front side or both front sides opens the *caddy* up; a CD disk then can be put inside the caddy. An open slot or a metal cover on the bottom slides out of the way to allow laser beam access to the CD by the drive. The *caddy* is inserted into the CD-ROM drive similar to a tape or a cartridge.

Of the two mechanisms, the tray is far more common because it makes for a cheaper drive and easier to load/unload the CD.

3. Single and Multiple Drives

Some PC systems allow the use or one or more *CD-ROM* drives at a time. This can give the PC a great flexibility, if one or two CD disks are used very frequently or to duplicate a CDs.

- *Connectors and Jumpers*

The connectors and jumpers on a *CD-ROM* are usually similar to the ones on hard disk drive ones. *CD-ROM* drives are somewhat standard in the use of jumpers and connectors, and even where they are located on the drive. There is a standard 4-pin power connector on the back of a regular internal *CD-ROM* drive, the same kind that is used for hard disk drives and most other internal devices. The other connections and jumpers depend on the interface that the drive is using; an *IDE/ATAPI* drive will use different ones than a *SCSI* drive. The *IDE/ATAPI* uses the standard 40-pin data connector, along with jumpers to select the drive as a master or slave device. The *SCSI* is a 50-pin connector along jumpers to set the device ID and termination. One connector that is found on a *CD-ROM* and not on a hard disk drive is the *audio connector* that goes to the sound card. This three- or four-wire cable is used to send CD audio output directly to the sound card so it can be recorded or played back on the computer's speakers.

- *Logic Board*

Every *CD-ROM* drive contains a *logic board*. Similar to the hard disk drive, the function of the *logic board* is both to control the drive and to interface to the PC either using *IDE/ATAPI* or *SCSI*.

- *Audio Output and Controls*

CD-ROM drives are usually built with convenient features to allowing them to play and listen to multimedia or audio CDs. CD drives vary in terms of what they provide on the front panels, but usually have some of the following:

  o *Stereo Headphone Output:* This is a mini headphone jack that allows headphones to be plugged directly into the drive and listen to the CD audio being played back.

  o *Volume Control Dial:* Most CD drives include a dial control to allow the volume level of the CD audio output to be set.

  o *Start and Stop Buttons:* Some drives include control buttons to start and stop the play of the CD. On some drives these are the only controls found on the front panel.

  o *Eject Button:* Most CD-ROM drive include a manual eject (*Tray* or *Caddy*) button.

- *Compact Disk Formats*

At the lowest level of recording, all CD record information is the same. On a spiral track using lands and pits, there are many different types of data that can be placed on a CD. For example, an audio CD contains bits and bytes just like a CD-ROM data CD, but the information is laid out in a totally different manner. These different ways of organizing the ones and zeros on the disk are called *CD formats*. There are several different formats in use today. Some are more popular than others; some require special drives to access them, while others are compatible with each other to some degree. These formats are basically equivalent to the logical structures and file systems used on hard disks or floppy disks. Unlike other media, CD formats are fixed in terms of how the data is structured, storage capacity and block size. There is no similarity to "*format*" a compact disk the way a hard disk or floppy disk is formatted, and there is no concept of partitioning either. The structures are basically the same for each CD that uses that particular format. The sheer number of different formats, and the fact that some drives will handle certain formats while others will not, makes all of this a rather confusing issue. The following is a description of the different formats:

o **Format Sessions:** *Sessions* of a CD is the number of different continuously written blocks of data that are placed on the disk. Traditional CD formats such as standard audio CD and data CD have *Single Session*; data on the CD is placed at once when the CD disk is manufactured. Some new CD formats however use more than one session, and are called *Multi Session* drives. Data information on these CD's can be written to the first part of the disk, and then later more information can be added to it in the unused space left after the first session. *Multi Session* CD capabilities are provided by many new drives. Some formats can use single or multiple sessions. For example, CD's that contains photo images may be a single session, but to add additional images to the CD disk requires a multi session capable drive. CD-Recordable (CD-R) disks can be either single or multi session CD drive. Standard audio CDs and data CDs are usually *Single Session*. The CD-Rewriteable (CD-RW) has Multi Session capabilities. CD disks written in CD-RW format can only be properly read by a drive that supports *Multi Session* disks.

o **Book color format:** CD media has many different formats. Each one of these formats has a formal name, but also the color of a book. This color is the initial paper manual describing the specification and standards. *Table 12.1* is a summary *Book Color* of CD formats:

Book Color Format	
Color	*CD Format*
Red	CD Digital Audio (CD-DA)
Yellow	Digital Data (CD-ROM), CD-ROM Extended Architecture (CD-ROM XA)
Green	CD-Interactive (CD-I)
White	Bridge CD (Photo CD, Video CD)
Orange	Magneto-Optical (MO), CD-Recordable (CD-R), CD-Rewriteable (CD-RW)

Table 12.1

Some of these books, especially yellow, refer to multiple formats, and also have format extensions associated with them. There may be some confusion in terminology because some of the formats are extensions to and derivations of earlier ones.

- **RED - Compact Disk Digital Audio (CD-DA)**

  The *RED CD* format was the first, which defined the audio CD used in all regular CD players, called *CD Digital Audio* or *CD-DA*. The specifications for this format were codified in the first CD standard, the so-called "*Red Book*" that was developed by Philips and Sony, the creators of the original compact disk technology. The "*Red Book*" was published in 1980, and actually specifies the data format for digital audio and the physical specifications for *Compact Disks*: the size of the media and the spacing of the tracks, for example. All of the subsequent CD standards that came after *CD-DA* built on the "*Red Book*" specification, because they use the same specifications for the media and how it is read. They also base their structure on the original structure created for audio CD. Data in the digital audio CD format is encoded by starting with a source sound file then sampling it to convert it to digital format. *CD-DA* audio uses a sample rate of 44.1 KHz (Kilohertz or 44,100 Hertz), which is about double the highest frequency audible by humans (approximately 22 KHz.) Each sample is 16-bit in size, and the sampling is in stereo sound,

$$44,100 * 2 * 2 = 176,400 \quad (Eq.\ 12.0)$$

Therefore, each second of sound takes 176,400 bytes of data. Audio data is stored on the disk in blocks, also called *sectors*. Each block holds 2,352 bytes of data, with an additional number of bytes used for error detection and correction, as well as control structures.

$$\frac{176,400}{2,352} = 75 \quad (Eq.\ 12.1)$$

Therefore, 75 blocks are required for each second of sound. On a standard 74-minute CD then, the total amount of storage is,

$$2,352 * 75 * 74 * 60 = 783,216,000 \quad (Eq.\ 12.2)$$
(NB: 60 seconds is 1 minute)

$$\frac{783,216,000}{1,048,576} = 746.9 \quad (Eq.\ 12.3)$$

From the 783,216,000 bytes *(12.2)* or about 747MB *(12.3)* derives the *Rule of Thumb* that a minute of audio CD is about 10 MB of data uncompressed. Using special software, it is possible to actually read the *Digitally Encoded* audio data directly from the CD and store it in a computer sound format such as a WAV file. However, while every CD-ROM drive will playback standard *"Red Book"* digital audio, every drive may not allow reading the CD data directly, also called *CD-DA extraction*. The reason may be that some drive manufacturers intentionally program the drives not to allow data extraction, in order to avoid possible copyright infringement by their owners.

- ### YELLOW - CD-ROM Digital Data (CD-ROM)
  This CD format is also referred to as *"ISO 9660"* or *"High Sierra"*. The standard that describes how digital data is to be recorded on a compact disk media went through several different iterations before the format was finalized. The first step was the creation of the format standard, the *"Yellow Book"*, by Philips and Sony in 1983. The specification was based on the original *"Red Book"* format. The *"Yellow Book"* specification was general enough that it was feared that many different companies would implement proprietary data storage formats using this specification, resulting in many different incompatible data CDs. To prevent this, major manufacturers met, in 1985, at the *High Sierra Hotel* and *Casino* in *Lake Tahoe, Nevada, USA*. They introduced a common standard format for data CDs. This format was named *High Sierra Format*. It was later modified slightly and adopted as an *"ISO 9660"* standard. The terms *"Yellow Book"*, *"High Sierra"* and *"ISO 9660"* are used somewhat interchangeably to refer to standard data CDs, although the most common name is

simply: "*CD-ROM*". Virtually all data CDs that are in use today are standardized and work in all standard *CD-ROM* drives. From the data CD standard, there are three modes defined:

1. ***Mode 1:*** This is the standard data storage mode used by virtually all standard data CDs. The data is written in basically the same way as it is in standard audio CD format "*Red Book*", except that the 2,352 bytes of data in each block are broken down into 2,048 bytes for "*real*" user data per sector, the other 304 bytes for an additional level of error detecting and correcting code. This is necessary because data CDs cannot tolerate the loss of bits the way audio CDs can.

2. ***Mode 2:*** This mode data CD is the same as *Mode 1* CD except that the error detecting and correcting codes are less. The reason is that *Mode 2* format provides a more flexible vehicle for storing more types of data that do not require high data integrity: for example, graphics and video can use this format. Furthermore, different kinds can be mixed together; this is the basis for the extensions to the original data CD standards known as "*CD-ROM Extended Architecture*", or "*CD-ROM XA*". Each block contains 2,048 bytes of "*real*" data. As with the audio format, there are 75 blocks per second of the disk. On a standard 74 minute compact disk, this yields a total capacity of,

$$2,048 * 75 * 74 * 60 = 681,984,000 \quad (Eq.\ 12.4)$$

$$\frac{681,984,000}{1,048,576} = 650.4 \quad (Eq.\ 12.5)$$

From 681,984,000 bytes, *(12.4),* derives the commonly known 650MB. Since the CD is designed to allow the reading of 75 blocks per second, this is the basis for the standard single-speed transfer rate of,

$$75 * 2,048 = 150\ \text{KBps} \quad (Eq.\ 12.6)$$

3. ***Mixed Mode:*** This mode is basically a mixture of data and audio information stored onto the CD, i.e., on track of data followed by audio tracks.

Similar to the way a hard disk or floppy disk has a file allocation table and root directory to identify the place to look in order to find the various directories and files on the disk, a data CD needs this "starting point" as well. At the start of the CD, a "*table of contents*" lists the contents of the disk and where to find it. New CD formats that are "*Multi Session*" can have more than one set of data on the disk, recorded at different times, and therefore use "*multiple tables of contents*", one per session. The *table of contents* is also called the *index of the disk*.

- ***CD-ROM Extended Architecture (CD-ROM XA):*** In 1988, the *CD-ROM Extended Architecture* or *CD-ROM XA* was developed by

Philips, Sony and Microsoft. This format built on the existing CD standard and was considered an extension to the original "*Yellow Book*" format. CDs that use CD-ROM XA can mix standard data CD *Mode 1* and *Mode 2* tracks, allowing the mixing of standard data along with other types of data. The *Mode 2* tracks are further divided into two types: *Form 1* and *Form 2*. Between all these different modes and forms, a CD-ROM XA disk can store data, audio, compressed audio, video, compressed video, graphics and others. The mixing together of these different types of information is called *interleaving*. Using a CD-ROM XA disk usually requires a drive specifically certified to be capable of reading the CD-ROM XA format. These drives may include hardware decoders to allow "*on-the-fly*" decompression of the compressed audio or video data for the capability of reading them. The drive usually handles the different data formats on the CD.

- **GREEN - CD-Interactive (CD-I)**

In 1986, Philips and Sony again joined forces to create the *CD-Interactive* or *CD-I* format. The goal was to develop both a format and a special new type of hardware to use it. In some ways this was the first attempt to create what is known as "*Multimedia*", with authors creating disks including text, graphics, audio, video, and computer programs, as well as hardware to handle all of these and be connected to a television screen for output. CD-I is derived from the original standard CD-ROM "*Yellow Book*" format, the way that *CD-ROM Extended Architecture (CD-ROM XA)* is. However, the format used by CD-I is somewhat different. A new class of disks called "Bridge Disks" was created that will work in drives subscribing to either format. The CD-I format was not implemented on many PC drives.

- **White - Bridge CD**

The term *Bridge CD* is used to refer to disks that use extensions or derivations of the *CD-ROM XA* format. These extended formats are described in the "*White Book*" specification. The reason for the term "*bridge*" is that these CD disks are designed to work in CD-ROM XA and CD-I drives, thus "bridging" the two formats of CD hardware. It is therefore possible to use CD-I disks in CD-ROM XA drives and CD-ROM XA disks in CD-I drives.

- **Video CD (VCD):** The support for a special CD format to store compressed video information is defined as part of the "*White Book*" specification. Through the use of *MPEG* compression it is possible to store 74 minutes of full-motion video in the same space that uncompressed "*Red Book*" audio uses. This format is called *video CD* or sometimes *VCD*. Playback video CDs requires either a video CD player or a CD-ROM drive that is video CD compatible. Since the compression algorithm used for video CD, *MPEG-1*, is unsophisticated, the quality of these disks has not been good.

- **Orange - Photo CD**

Developed in the early 90s by *Kodak* and *Philips*, *Photo CD* is an implementation of *CD-ROM XA* designed to hold photographic images. They technically use *Mode 2 Form 1* of the CD-ROM XA architecture. *Photo CD* is defined in the "*Orange Book*" specification, it is a multi-session discs that can be written incrementally. *A session is a set of data written in a single setting.* When a film for processing to photo CD is

121

sent, the film is first developed normally. The developed and printed pictures are then scanned and converted to digital form, encoded into the photo CD format, and written to the CD. Writing the photos to the CD is done using a process that is basically the same as how *CD-R* works: a laser burns the information into the tracks of the CD. After film is sent and created by the first photo CD, it is possible to record additional films to the same disk. However, doing this means that the information is written in *Multiple Sessions*, and therefore a drive that supports multiple sessions is required to access the disk properly. A photo CD can be written so that it is a *"bridge CD"*, which will allow it to be read by *CD-I* drives as well. *Part I* of the *"Orange Book"* defines the specifications for *Magneto Optical (MO)* drives. *Part II* of the *"Orange Book"*, published by Philips, specified the characteristics and format of a recordable CD, the *CD-Recordable* or *CD-R. CD-R* is also called *CD-WORM (Write-Once, Read-Many)* or *CD-WO (WO* means *"Write Once").*

- *Magneto-Optical disk drive (MO):* The *magneto-optical (MO)* diskette/disk drive is a popular way to back up files on a PC. A *MO* device employs both magnetic and optical technologies to obtain ultra-high data density. A typical *MO* cartridge is slightly larger than a conventional 3½ inch magnetic diskette, and looks similar. But while the older type of magnetic diskette can store 1.44MB of data, the *MO* diskette can store many times that amount, ranging from 100 MB up to several Gigabytes. An *MO* system achieves its high data density by using a laser and a magnetic read/write head in combination. Both the laser and the magnet are used to write data onto the diskette. The laser heats up the diskette surface so it can be easily magnetized, and also to allow the region of magnetization to be precisely located and confined. A less intense laser is used to read data from the diskette. Data can be erased and/or overwritten an unlimited number of times, as with a conventional 3½ inch diskette. Examples of magneto-optical drives are the *Fujitsu DynaMO*, a 230 MB drive used in the *PowerPC Apple Powerbook*, a note book computer, and the Pinnacle Micro Vertex, a 2.6GB drive. The assets of *MO* drives include convenience, modest cost, reliability, and (for some models) widespread availability approaching industry standardization. The limitation of *MO* drives is that they are slower than hard disk drives, although they are usually faster than conventional 3½ inch diskette drives.
- *CD-Recordable (CD-R):* The *CD-R* was introduced in 1993. It replaced the conventional CD metal disk with a thin gold of data or about 24 *CD-ROMs. CD-R* drives allow the PC to create an audio or a data CD in various formats that can be read by most standard CD players or *CD-ROM* drives. As *"Write Once"* implies, the disk starts out blank, can be recorded once, and thereafter is permanent and cannot be re-recorded. *CD-R* is more than just another standard CD format. The *CD-ROM, CD-ROM XA, CD-I* drives were built upon the original audio CD standard by simply changing the interpretation of what the bytes in the original audio CD format meant. *CD-R* is actually in many ways the opposite: it defines new physical media and ways of recording them, while continuing to use the standard formats defined in other specifications. *CD-R* can be useful in several applications, including archiving, software distribution, backup and custom audio.

The reason that original CD disk was not able to record data is because the ones and zeros were encoded using physical changes to the CD disk: *pits* that are physically etched into the plastic substrate. *CD-R* technology was faced with the difficult task of finding a way to conveniently create these pits, without the special equipment required when mastering a CD disk. *CD-R* media addressed this issue by eliminating pits and lands entirely and using a different kind of media. *CD-R* media starts with a polycarbonate substrate, just like regular CDs do. Instead of physical etching this substrate, it is stamped with a spiral pre-groove, similar to the spiral found on a regular CD except that it is intentionally "*wobbled*". This groove is what the *CD-R* drive uses to follow the data path of the disk during recording. If the disk were totally blank then writing the spiral tracks would be a very complex process. On top of the polycarbonate, a special photosensitive dye layer is deposited; on top of that a metal reflective layer is applied (such as a gold or silver alloy) and then finally, a plastic protective layer. It is these different layers that give CD-R media their different visual appearance from regular CDs. The key to the media is the dye layer and the special laser used in the drives. It is chosen so that it has the property that when light from a specific type and intensity of laser is applied to it, it heats up rapidly and changes its chemical composition (during recording or "*burning*" of a CD disk). As a result of this change in chemical composition, the area "burned" reflects less light than the areas that do not have the laser applied. This system is designed to mimic the way light reflects cleanly off a "*land*" on a regular CD, but is scattered by a "*pit*", so an entire disk is created from burned and non-burned areas, similar to how a regular CD is created from pits and lands. The result is that the created CD media will play in regular CD players as if it were a regular CD, in most cases. From a process of heat and chemistry, the change is physically altering the CD media surface to a permanent and irreversible medium. Once any part of the CD has been written, the data is there permanently. Some drives allow recording information in a *Single Session* or a *Multi Session* or both. This, in turn, requires a CD player capable of recognizing a *Multi Session* disk in order to use the burned disk. *CD-R* media, generally, is 74 minutes in length (some disks are with more or less capacity) and about 650 MB.

Just as CD-R requires the use of special media, it also requires the use of a special CD-R drive. This drive is very different than a standard CD player because it must include a special laser. The laser is the key component from the drive's perspective, in that it is what burns the image into the CD-R media's dye layer. CD-R drives are capable of reading disks as well as writing them. Most drives support a large number of formats, to enable the reading and writing of a wide variety of CDs. They are also faster than single speed; it is typical for the speed of the drive to be significantly lower when writing than when reading. For example, a drive might be specified to be 4X when reading, but only 2X when writing a CD disk. While standard CD-ROM drives often use a variety of formats, SCSI is the interface of choice for the vast majority of CD-R drives. The main reason is that SCSI is a higher-performance interface that allows the flow of data to the drive to be maintained more easily and more independently of what other activities are happening within the PC. This is critical for writing *CD-R* media because burning a disk requires an "uninterrupted flow of data" from wherever it is coming from (usually a hard disk) to the drive. *CD-ROM* drives are also available with the *ATAPI (IDE)* interface. Since the *CD-ROM's* laser is moving at a constant speed as it writes, it must have the data it needs available in a smooth flow; it cannot wait for the data if it is delayed because the disk cannot be stopped. The faster the drive spins the disk when writing, the more data flow is required. Due to how CDs are written, interrupting the writing of the disk generally ruins it. Burned disks that do not work are referred to as "coasters". Many drives take specific steps to avoid this problem. One of the most common steps is the use of a substantial memory buffer that can supply data to the laser head in the event that the flow of data is interrupted. Another is the use of an image file. Instead of copying the data to be recorded from its original locations an exact image of the disk to be created is stored in a large file on disk first, and then transferred "whole" to the burned disk. This reduces the chance of an interruption while writing, but costs a good chunk of disk space. Many users have a hard time when initially setting up their *CD-R* drives, getting the drives to write disks properly without destroying too much blank media in the process. Once the drives are set up, however, *CD-R* can be a fairly reliable format, allowing the user to create a large number of disks with minimal waste.

CD-R drives require special software and usually is it supplied with the drive. This software is used not only to control the creation of a CD-R disk, but also to allow the user to compose and arrange the disk. The software will allow a user to create a custom audio CD (arrange the tracks in a preferred order). It will also allow the organization and management of mixed-mode disks containing multiple formats, as well as the creation of an image file for better performance in burning CDs. The software has a direct impact on how easy it is to make CD disks and can be the difference between success and failure in many cases.

The disks created by most *CD-R* drives are compatible with most *CD-ROM* and audio CD disks. Some old CD drives can have problems with the output of some *CD-R* drives. Some players can actually have more problems with some types of media than with others, since they use different dye and/or reflective metal layers. Changing the type of *CD-R* blanks used can sometimes fix an incompatibility problem. The drive itself also has a big impact on the readability of *CD-R* it creates. *CD-R* disks that are written in multi-session format (writing part of the disk at one time and another part later on) can only be read in a drive that is multi-session capable.

- ***CD-ReWriteable (CD-RW):*** *Part III* of the *"Orange Book"* defines rewriteable CDs, which are erasable CD-R. A new technology, called *CD Rewriteable* or *CD-RW* is also referred to as *Erasable CD* or *CD-E*. *CD-R* moved beyond the read-only standard compact disk by allowing the writing of CDs by the PC user. It however has its limitations as well, chiefly the fact that each disk can only be written once. *CD-RW* enables the CD to be both written and rewritten.

CD-RW uses a more advanced technology in order to accomplish its goal of making compact disks both writeable and rewriteable. There are therefore many issues related to operation of these drives, and also compatibility, that are not applicable to most other CD formats.

CD-RW media is more similar to *CD-R* media than to the original *CD-ROMs*. *CD-R* media replaces the physically molded pits in the surface of the disk with a sensitive dye layer that can be burned to simulate how a standard CD works. *CD-RW* media is formed in the same basic concept as the *CD-R* media; it starts with a polycarbonate base and a molded spiral pre-groove to provide a base for recording. There are several layers applied to the surface of the disk, with one of them being the recording layer where ones and zeroes are encoded. The recording layer for *CD-RW* is different than the *CD-R*: the dye layer used on the *CD-R* is permanently changed during the writing process, which prevents rewriting. *CD-RW* media replaces this dye with a special *phase-change* recording layer, comprised of a specific chemical compound that can change states when energy is applied to it, and can also change back again. Depending on its temperature, there are some types of chemicals that can not only change their state after having heat or other conditions applied, but also even retain that state when the heat is removed. They can later be returned to their original state through another, different process. The material used in *CD-RW* disks has the property that when it is heated to one temperature and then cooled, it will crystallize, while if it is heated to a higher temperature and then cools, it will form a non-crystalline structure when cooled. When the material is crystalline, it reflects more light than when it is not; so in the crystalline state it is like a "*land*" and in the non-crystalline state, a "*pit*". By using two different laser power settings, it is possible to change the material from one state to another, allowing the rewriting of the disk. The change of phase at each point on the disk's spiral is what encodes ones and zeros into the disk. The spiral and other structures are the same as for *CD-R*; what changes is how the pits are encoded. *CD-RW* media have one very important drawback: they do not emulate the pits and lands of a regular CD as well as the dye layer of a regular *CD-R*, and therefore, they are not backward compatible to all regular audio CD players and *CD-ROM* drives. In addition, the fact that they are written multiple times means that they are multi-session disks by definition, and so are not compatible with non-multi-session-capable drives.

CD-RW drives are similar to *CD-R* drives except that they employ a very different kind of laser, to enable them to write the special *CD-RW* media. Like *CD-R* drives, *CD-RW* drives are capable of writing in multiple different standard formats, and reading those formats as well. *CD-RW* drives can also write *CD-R* media, making them extremely flexible. There is one significant disadvantage to *CD-RW* drives, however, compared to *CD-R* drives: they are basically limited to writing at 2X speeds. Unlike *CD-R*, the actual writing mechanism is tied to a specific speed, which is currently 2X as specified in the *"Orange Book"* standard. In order to create 4X *CD-RW*, for example, a new standard would need to be created.

- *Digital Versatile/Video Disk (DVD):* To provide for the next generation of software, a new format has been created called *Digital Video Disk or Digital Versatile Disk or DVD. DVD* uses the same physical form factor as CD-ROM but the logical formats are considerably different. A typical *DVD* can store up to 17GB of data or about 24 CD-ROMs. It is Read-Only Media. A new version of *DVD* called *DVD-RAM* may even someday replace the home VCR. It is essentially a bigger, faster CD that can hold video as well as audio and computer data. *DVD* aims to encompass home entertainment, computers, and business information with a single digital format, eventually replacing audio CD, *VCR, Laserdisc, CD-ROM,* and perhaps even video game cartridges. *DVD* has widespread support from all major electronics companies, all major computer hardware companies, and some of the major movie and music studios. It is important to understand the difference between *DVD-Video* and *DVD-ROM. DVD-Video* (called *DVD)* holds video programs and is played back on a *DVD* player connected to a TV. *DVD-ROM* holds computer data and is read by a *DVD-ROM* drive connected to a computer. The difference is similar to that between Audio CD and CD-ROM. *DVD-ROM* also includes recordable variations *(DVD-R, DVD-RAM, DVD-RW, DVD+RW)*. Some new computers with *DVD-ROM* drives can also play *DVD-Videos* but not the other way around. *DVD* is an optical disk technology that is expected to rapidly replace the CD-ROM disk (as well as the audio compact disc) over the next few years. The digital versatile disk *(DVD)* holds 4.7GB of information on one of its two sides, or enough for a 133-minute movie. With two layers on each of its two sides, it will hold up to 17GB of video, audio, or other information (compare this to the current CD-ROM disk of the same physical size, holding 600MB, the *DVD* can hold more than 28 times as much information). *DVD-Video* is the usual name for the *DVD* format designed for full-length movies and is a box that will work with the television set. *DVD-ROM* is the name of the player that will replace the computer's CD-ROM. It will play regular CD-ROM disks as well as *DVD-ROM* disks. *DVD-RAM* is the writeable version. *DVD-Audio* is a player designed to replace the compact disc player. *DVD* uses the *MPEG-2* file and compression standard. *MPEG-2* images have four times the resolution of *MPEG-1* images and can be delivered at 60 interlaced fields per second where two fields constitute one image frame. *MPEG-1* can deliver 30 noninterlaced frames per second. Audio quality on DVD is comparable to that of current audio compact disks. *DVD* is a new type of CD-ROM that holds a minimum of 4.7GB, enough for a full-length movie. Many experts believe that *DVD* disks, called *DVD-ROMs,* will eventually replace CD-ROMs, as well as VHS video cassettes and laser discs. The *DVD* specification supports disks with capacities of from 4.7GB to 17GB and access rates of 600 KBps to 1.3 MBps. One of the best features of DVD drives is that they are backward-compatible with CD-ROMs. This means that DVD players can play old CD-ROMs, CD-I disks, and video CDs, as well as the new DVD-ROMs. New *DVD* players, called *second-generation* or *DVD-2 drives,* can also read CD-R and *CD-RW* disks. *DVD* uses *MPEG-2* to compress video data.

o **DVD+RW:** *DVD+RW* is a new standard for rewritable *DVD* disks being promoted by Hewlett-Packard, Philips and Sony. The *DVD-RAM*, which was developed by the *DVD* Consortium, and the *DVD+RW* standards are incompatible. *DVD+RW* disks have a capacity of 3GB per side.

o **DVD RAM:** *DVD RAM* is a new type of rewritable compact disc that provides much greater data storage than today's CD-RW systems. The specifications for *DVD-RAMs* are still being defined by the *DVD* Consortium. The *DVD-RAM* standard supports 2.6GB per disk side.

o **DVD-Video:** *DVD-Video* is a video format for displaying full-length digital movies. A number of manufacturers are just beginning to produce *DVD-Video* players, which attach to a television just like a videocassette player. Unlike *DVD-ROMs,* the Digital-Video format includes a *Content Scrambling System (CSS)* to prevent users from copying discs. This means that *DVD-ROM* players cannot play *DVD-Video* discs without a software or hardware upgrade to decode the encrypted discs.

o **Digital Video Interactive (DVI):** *DVI* is a technology developed by General Electric that enables a computer to store and display moving video images like those on television. The most difficult aspect of displaying TV-like images on a computer is overcoming the fact that each frame requires an immense amount of storage. A single frame can require up to 2MB of storage. Televisions display 30 frames per second, which can quickly exhaust a computer's mass storage resources. It is also difficult to transfer so much data to a display screen at a rate of 30 frames per second. *DVI* overcomes these problems by using specialized processors to compress and decompress the data. *DVI* is a hardware only *codec (compression/decompression)* technology. A competing hardware codec, which has become much more popular, is *MPEG. Intel* has developed a software version of the *DVI* algorithms, which it markets under the name *Indeo.*

o **Digital Video Express (Divx):** *Divx* is a new *DVD-ROM* format being promoted by several large Hollywood companies, including *Disney, Dreamworks SKG, Paramount* and *Universal.* With *Divx,* a movie (or other data) loaded onto a *DVD-ROM* is playable only during a specific time frame, typically two days. As soon as it begins playing a *Divx* disc, the counter starts. Each *Divx* player is connected to a telephone outlet and communicates with a central server to exchange billing information. *Divx* discs have the potential to ultimately replace video tapes. They are especially convenient for video rentals because there are no late fees. A *Divx* title can be purchased to keep. However, *Divx* format is not backward-compatible with current *DVD-ROM* players. This means a *Divx* player only play *Divx* titles.

- **CD Speed, Constant Linear Velocity (CLV) and Constant Angular Velocity (CAV)**

Every computer's device or component has a metric: Processors are the MHz clock speed; Memory the *DRAM* chips speed; Monitors are often the nominal screen size; and CD-ROM drives are the "*X*" speed. The "*X*" speed of the drive refers to its nominal spin rate. A 1X drive spins that the speed of a standard audio CD, a 2X drive spins at twice this speed, etc. Since the CD-ROM drive uses *Constant Linear Velocity (CLV)*, this means that the actual spin speed of the disk changes from about 210 to 539 RPM depending on whether the inside or outside of the disk is being read. A 2X drive would double this to a range of 420 to 1078 RPM, etc. Most drives (12X and down) use *CLV*. New and faster drives changed to a fixed spindle speed, which is a system called *Constant Angular Velocity* (*CAV*). The spindle speed, in a *CAV* drive, remains the same and the transfer rate changes depending on where the disk is being read.

The performance of a CD drive does not scale *linearly* with the "*X*" number of the drive. For example, a 12X drive will not transfer data at 12 times the speed of a 1X drive, and will not have 1/12th the access time, etc. With a standard *CLV* drive, the "*X*" number refers to the spin rate and therefore the theoretical maximum transfer rate is

Theoretical transfer rate = 12 * 150 KB/s = 1.8 MB/s. *(Eq. 12.7)*

This is only *theoretical* however, and is affected by command overhead, the interface speed, memory, etc. The access time of the drive, which refers to how quickly it can move around on the disk when not reading sequentially, gets better with faster drives but definitely *not linearly*. With new CAV drives, the "*X*" cannot be used to estimate the transfer rate. Since the drive spins at the same speed but there is less data in the middle of the disk, the highest transfer rate only applies to reading data from the outermost part of the disk. A 24X drive is only 24X at the edge of the CD; in the middle it might be significantly slower (as much as 60%). On a CD, typically the data is recorded on the inside first, and then moves to the outside. If the disk is half-full then, none of the data will be read back at anything even near 24X on such a drive. On the other hand, the *CAV* drives also have the advantage of not having to change speeds, which can make them perform more smoothly even if their raw transfer rate is below that implied by the theoretical number. It is also important to realize that not all drives with the same "*X*" rating are created equal. They may all spin at the same speed, but differences in control circuitry or the ability to speed up or slow down the motor can mean that one brand name 4X drive performs a lot better than another brand.

- **Speed Change Time:** One performance factor that is relevant for some CD-ROM drives is *Speed Change Time*. A typical CD-ROM drive uses *CLV* reading, which means that the same amount of data is read from the disk in each unit of time. This means the speed of the disk must change as the head is moved from the inner part of the disk to the outer, or vice-versa. It can take a significant amount of time to change the speed of the spindle motor, especially with faster drives. The difference in speed can be several thousand *Rotation Per Minute* or *(RPM)* in the faster drives, between the speed at the inside and the outside of the disk. When the disk is doing a lot of random accesses, noise can be heard from the motor audibly; *"ramps up"* in frequency and *"ramps down"* as it moves around the disk. The delay in changing speeds can slow down random accesses on the disk, sometimes significantly. This speed change rate is not usually specified directly by the manufacturer, but instead is usually lumped in to the access time measurement. Changing the speed of the disk can be done at the same time as the head is moved during a seek. If the seek is completed, however, and the drive has not finished changing speeds yet, there will be a small delay. Due to the amount of time it takes to change speeds, many new drives have changed to *CAV*. These drives do not change speeds, and therefore do not incur the speed change penalty as they move over the disk. However, they give up some transfer rate performance as they move to the inside of the disk, since they are covering less linear distance in each unit of time. A 24X CD-ROM drive is only 24X at the outer edge of the disk, but it may be the equivalent of at least 12X even on the inside of the disk.

- ***Seek Time:*** *The seek time of a drive refers to the amount of time that it takes to move the heads to a specific part of the disk in order to do a read.* The amount "of time that it takes" depends on how far away from the destination the heads are; in most cases seek time is measured as an average for a typical random read on the surface of the disk. In actual fact seek time as a metric is used for indicating the performance level of hard disks much more often than it is for CD-ROM drives; for CDs it is much more common to see the access time metrically stated, of which seek time is a component. Overall, CD-ROM drives have a poor seek time performance than hard disks do. They use less efficient head actuator mechanism that causes them to take much more time to position to different tracks on the surface of the disk. This is probably because of the legacy of CD-ROM technology, which started out with audio CD.

- ***Latency:*** The CD media is a spinning disk of information. When a particular part of the disk needs to be read, the heads must be moved to the correct part of the disk; the speed of this action is measured by *seek time*. Once the head is in the correct place, it will start reading the track of information, but may have to wait for the correct data to turn around on the disk and reach where the head is, since the CD is spinning. *The amount of time it takes, on average, for the correct information to come around to where the head is waiting after a seek is referred to as latency.* Since latency measures the amount of time that it is taking for the disk to spin into the right position, latency is directly correlated to the speed that the disk is spinning. This means that drives with a faster "X" rating will have lower latency, because they are spinning faster. This is one reason why these drives will have better performance. Measuring latency is more complicated for standard CD-ROM drives than it is for hard disks because hard disks are spinning at a constant speed while conventional CD-ROM drives are not. While new CD-ROM drives use *CAV* and spin at the same speed all the time, conventional CD-ROMs use *CLV* and spin faster when reading the outside of the disk than when reading the inside. This means that these drives will have better latency performance when reading the outside of the disk. As with seek time, latency numbers are rarely seen for CD-ROM drives. Instead, access time is used as an overall indicator of the amount of time to access a random part of a CD in the drive. Latency is only one component of access time, which is really the measure to look at in more detail when considering random-access performance of a CD-ROM drive.

- *Access Time:* One of the most common performance statistics for CD-ROM drives is *access time*. With hard drives it is much more common to see quotes of the other metrics that are combined to make up access time. *Access time is the amount of time it takes from the start of a random read operation until the data starts to be read from the disk.* It is a composite metric, really being composed of the following other metrics:

  o **Speed Change Time:** For *CLV* drives, the time for the spindle motor to change to the correct speed.

  o **Seek Time:** The time for the drive to move the heads to the right location on the disk.

  o **Latency:** The amount of time for the disk to turn so that the right information spins under the read head.

  Although access time is made up of the time for these separate operations, adding these measurements together would not get access time. The relationship is more complex than this because some of these items can happen in parallel. For example, the speed of the spindle motor may vary at the same time that the heads are moved. The access time of CD-ROM drives in general depends on the rated "X" speed of the drive, although this can vary widely from drive to drive. The old 1X CD drives generally had an access times exceeding 300 milliseconds (ms); as drives have become faster and faster, access times has decreased to below 100 ms on some drives. Note that while faster "X" rated drives have lower access times, this is due to improvements that reduce the three metrics listed above that contribute to access time. Some of it (latency for example) is reduced when the drive spins the disk at 8X instead of 1X. On the other hand, seek time improvement is independent of the spin speed of the disk, which is why some 8X drives will have much better access time performance than other 8X drives. Even the fastest CD-ROM drives are significantly slower than even the slowest hard disks; access time on some CD-ROM is about four or five times higher than that of some disk drive. CD-ROM drives are based on technology originally developed for playing audio CDs, where random seek performance is very unimportant. CDs do not have cylinders like a hard disk platter, but rather a long continuous spiral of bits, which makes finding specific pieces of data much more difficult.

- *Transfer Rate: The transfer rate of a drive refers to how quickly data can be moved from the surface of the CD and into the computer.* This is the primary statistic that measures how efficiently the PC can actually retrieve the data of the surface of the compact disk when it is needed. Since many applications involving CDs require the transfer of large blocks of data, the transfer rate of the drive is an important performance metric. There are two different factors that make up transfer rate:

  o *Internal Transfer Rate:* It measures the speed at which data can be actually read from the disk and into the CD-ROM drive's internal controller. The *Internal Transfer Rate* is the "true" transfer rate of the drive, since this is the rate that data can actually be read from the disk.

  o *External Transfer Rate:* It measures the speed at which data can be moved from the control over the CD-ROM interface and into the rest of the PC. The *External Transfer*

Rate is rarely an important factor with CD-ROM drives (unlike hard disks) because their internal transfer rates are so low that even slower interfaces are rarely a limiting factor. The theoretical maximum transfer rate of a 12X drive for example is only 1,800 KB per second, and most *SCSI* or *IDE/ATAPI* interface can handle this with no difficulty. The *External Transfer Rate* is more of a limiting factor; the data cannot be fed to the PC any faster than the interface's transfer rate, but having a faster interface is not good if the drive itself is slow.

Unlike hard disks, where far too much focus is on the external or interface transfer rate, most manufacturers of CD-ROMs quote the true internal transfer rate of the drive. The quoted specifications are always for ideal, peak transfers, and will never be achieved over a period of time in the "real world", due to overhead and other electronic circuit imperfections. Also, the real world is a mix of random accesses and sequential transfers, and the transfer rate measures only sequential transfers. Conventional CD-ROM drives use *CLV*. Since there is more data at the outside of the disk than at the inside, they make the disk spin slower when reading the outside of the disk so that the transfer rate over the entire surface of the disk is identical. This means that the theoretical *Internal Transfer Rate* of the drive can be calculated easily by looking at the "X" rating of the drive and multiplying it by 150 KB/second, the transfer rate for a standard 1X drive. This will be the transfer rate for an entire CD, no matter what part of it is being used.

The new *CAV CD-ROM* drives do not vary the speed of the disk depending on where they are reading. They spin the disk at the same speed all the time. This means that when reading the outside of the disk, where there is more data per revolution, they will have a higher transfer rate than when they are reading the inside of the disk (same as for hard disks). The difference is substantial because the ratio of the data density of the outside of the disk to the inside is about 2.5 to 1. The CD drive manufacturers only state in the specifications the theoretical transfer number, the transfer rate when reading the outside of the disk. Since a CD has its data recorded from the inside out, this means on a disk that is half-full, none of the data will be read back at 24X when using a *CAV* 24X drive. The data, at the very inside edge, will be read at the equivalent of less than a 12X drive. *Table 12.2* shows a summary of the transfer rates of different types of drives:

Tranfer Rate of a CD		
Drive	Minimum Transfer Rate	Maximum Transfer Rate (binary KB/s)
1X (CLV)	150 KB/s	150 KB/s
2X (CLV)	300 KB/s	300 KB/s
4X (CLV)	600 KB/s	600 KB/s
6X (CLV)	900 KB/s	900 KB/s
8X (CLV)	1,200 KB/s	1,200 KB/s
10X (CLV)	1,500 KB/s	1,500 KB/s
12X (CLV)	1,800 KB/s	1,800 KB/s
16X (CAV)	~ 930 KB/s	2,400 KB/s
20X (CAV)	~ 1,170 KB/s	3,000 KB/s
24X (CAV)	~ 1,400 KB/s	3,600 KB/s
12X/20X (CLV/CAV)	1,800 KB/s	3,000 KB/s

Table 12.2

CHAPTER XIII
MULTIMEDIA DEVICES

- ### *Multimedia*

   Multimedia is the use of computers to present text, graphics, video, animation, and sound in an integrated way. Until the mid 90's, multimedia applications were not common due to the expensive cost of hardware required (mainly CD ROM and sound card) and the computer's overall performance. With increases in performance and decreases in price, however, multimedia is now desirable on a PC. Since the mid 90's, nearly all PCs are capable of displaying video, though the resolution available depends on the power of the computer's video adapter and CPU. Because of the storage demands of multimedia applications, the most effective media are CD-ROMs. Multimedia has been part of the computer's CPU. In 1997, Intel introduced a microprocessor with the suffix "*MMX*". That was the first generation of Personal Computers with multimedia capabilities. A set of 57 multimedia instructions built into Intel's microprocessors, those *MMX*-enabled microprocessors can handle many common multimedia operations, such as *Digital Signal Processing (DSP)*, that are normally handled by a separate sound or video card.

- ### *Sound Card Specifications*

   Sound cards are adapter cards (*Figure 13.0*) that generate sound from the PC either through a program (wave file) or transfer (CD ROM). Sound card can convert input signals or information into simultaneous voices. For example, available sound cards such as the *Soundblaster 16*, Orchid *Soundwave 32*, and *Soundblaster AWE64* can output 16, 32 and 64 simultaneous *voices* respectively. These numbers do not refer to the bit resolution. Each *voice* represents a different, distinct instrument being played by the sound card; i.e., the piano sound playing would be one voice, the clarinet a second voice, and so on.

Figure 13.0

133

- ***Sample Rate***

The key in defining the characteristic of any digital audio signal is its *sampling rate*, or the *sample rate*. This refers to how frequently the *analog signal* is measured during the sampling process. An example of an analog signal is the human voice. The process of recording that voice (sound) and sampling the recording once per second is called the sample rate. The more frequently the signal is sampled, the better the approximation to the original sound. However, the higher the sample rate the more memory is required to store the samples.

A single sine wave (*Figure 13.1*) has one peak and one valley: one *cycle*. Imagine that the *sine wave* (or *waveform*) is a sound that occurs in one second of time, then the frequency of that sine wave is one cycle per second; the unit of measure for frequencies is *Hertz* or *Hz*, so that sine wave is 1 *Hz*. All sounds are composed of mixtures of sine waves of different frequencies. High-pitched sounds have higher frequencies, and low-pitched sounds lower frequencies. *The minimum sampling rate to reproduce a signal is twice the waveform's original frequency.*

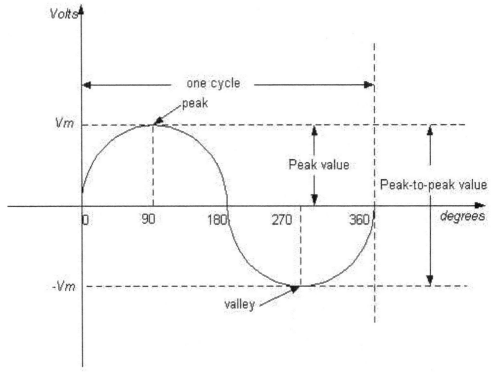

Human voice ranges from about 50 Hz up to about 4 kHz (4,000 Hz); the telephone, built to carry human voice, was designed to cover up to 8 KHz. The range of some musical instruments can reach 20 KHz, which is generally accepted as the highest frequency that human ears can discern. Therefore, to be able to capture these highest frequencies accurately, one must sample at least at double that rate. Compact disks are recorded at a sampling rate of 44.1 KHz. Sample rate can also be seen in other media as well. The television is a representation of real-time motion (analog) in a discrete form. Standard television is 30 *frames per second* (*fps*). That means every second, 30 pictures (frames or snapshots) of the original scene are displayed on the screen. For the most part consecutive pictures are the same, except for very small differences. A hand is moving, a racecar is coming down the track, a bullet is shooting out of a gun etc.. They are all single pictures put together end after end at a speed that the human eye perceives as a fluid motion. Now contrast this with older computer generated video formats (for example *AVI* files).

Generally they were recorded at 15 *fps*. Although this is fast enough to provide a reasonable semblance of motion, it does not compare to the 30 *fps* of the TV. Motion picture movies are actually recorded at only 24 *fps*.

- **Sample Resolution**

In one key determinant, *sample rate* affects the quality of digital audio. The other determinant of audio quality is the *sample resolution*. This refers to how many different values the samples can take on. The higher the sample resolution, the more accurate the representation of the level of each sample, but the more memory is required to store each sample. In digital sampling, each sample is represented by one of a limited number of discrete values. Digital audio is normally in one of two resolutions:

- o **8-Bit:** The 8-bit resolution was used in the earliest sound cards, and is used for some lower-quality recording formats as well. Each sample can take one of 256 different values ($2^8 = 256$). This is not generally considered enough resolution to accurately represent music audio.

- o **16-Bit:** This is the standard for CD audio and new sound cards. Each sample can take one of 65,536 different values ($2^{16} = 65,536$), more than any human can readily discern.

The 256 and 65,536 values are *per channel*. For *stereo* audio, which is high quality music, there are two channels (left and right). This doubles the sampling and memory requirements.

- **Playback and Sound Quality**

The digital audio samples produce must be converted to an analog signal for the human ear to hear. The process is now reversed and converted the 1's and 0's encoded on the CD or in a digital audio file to analog voltages to be sent to PC's speakers. The device that processes this conversion is called the *Digital to Analog Converter*, or *DAC (Figure 13.2)*. In the home stereo system, and in some sound cards, the *DAC* is very high quality. Typically, the lower the price of sound card, the lesser the quality. The quality of sound could also be related to the quality of speakers. Another hidden sound reproduction that may be a major problem on the PC is *noise*. Some computer components generate a *lot* of noise, such as electrical noise that can become audible if it is processed by the PC sound card circuitry. The power supply, hard drives, CD-ROM, tape backup, processor cooling fan, ZIP drive, all generate electrical noise. Some of this bleeds over into the sound card circuitry. It is recommended to install the sound card in the slot farthest from all of these devices.

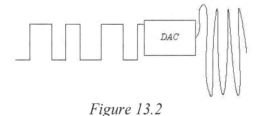

Figure 13.2

Digital audio is typically created by taking 16 bit samples a second of the analog signal. Since this signal is typically spread out over a spectrum of 44,100 cycles per second (44.1 *KHz*), this means that one second of CD quality sound requires 1.4 million bits of data.

Using their knowledge of how people actually perceive sound, the developers of *MP3* devised a compression algorithm that reduces data about sound that most listeners cannot perceive. MP3 is currently the most powerful algorithm in a series of audio encoding standards developed under the sponsorship of the *Motion Picture Experts Group (MPEG)* and formalized by the *International Organization for Standardization (ISO)*.

- *Moving Picture Experts Group (MPEG)*

The *MPEG* standards are an evolving set of standards for video and audio compression developed by an *ISO* group: the *Moving Picture Experts Group (MPEG)*. *MPEG* refers to the family of digital video compression standards and file formats developed by the ISO group. *MPEG* generally produces a better-quality video than competing formats, such as Video for Windows, Indeo and QuickTime. *MPEG* files can be decoded by special hardware or by software. It achieves high compression rate by storing only the changes from one frame to another, instead of each entire frame. The video information is then encoded using a technique called *DCT*. MPEG uses a type of *lossy compression*, since some data is removed. But the diminishment of data is generally imperceptible to the human eye. There are two major MPEG standards:

MPEG-1 implementations provide a video resolution of 352-by-240 at 30 *frames per second* (*fps*). This produces video quality slightly below the quality of conventional VCR videos. MPEG-1 was designed for coding progressive video at a transmission rate of about 1.5 million bits per second. It was designed specifically for Video-CD and CD-I media. MPEG-1 audio layer-3 (MP3) has also evolved from early MPEG work.

MPEG-2 offers resolutions of 720x480 and 1280x720 at 60 *fps*, with a full CD-quality audio. This is sufficient for all the major TV standards, including *NTSC*, and even *HDTV*. *MPEG-2* is used by DVD-ROM. It can compress a 2 hour video into a few Gigabytes. While decompressing an *MPEG-2* data stream requires a modest computing power, encoding video in *MPEG-2* format requires significantly more processing power. MPEG-2 was designed for coding interlaced images at transmission rates above 4 million bits per second. MPEG-2 is used for digital TV broadcast and DVD. An MPEG-2 player can handle MPEG-1 data as well.

A proposed MPEG-3 standard, intended for *High Definition TV (HDTV)*, was merged with the MPEG-2 standard when it became apparent that the MPEG-2 standard met the *HDTV* requirements. An MPEG-4 standard is a much more ambitious standard and addresses speech and video synthesis, fractal geometry, computer visualization, and an artificial intelligence (AI) approach to reconstructing images.

- o *Discrete Cosine Transform (DCT):* DCT is a technique for representing waveform data as a weighted *sum of cosines*. DCT is commonly used for data compression, as in *JPEG*. This usage of DCT results in lossy compression. *DCT* itself does not lose data; rather, data compression technologies that rely on *DCT* approximate some of the coefficients to reduce the amount of data.

o **Data compression:** *Data compression* is storing data in a format that requires less space than usual. *Compressing* data is the same as *packing* data. Data compression is particularly useful in communications because it enables devices to transmit the same amount of data in fewer bits. There are a variety of data compression techniques, but only a few have been standardized. The CCITT has defined a standard data compression technique for transmitting faxes (Group 3 standard) and a compression standard for data communications through modems (CCITT V.42*bis*). In addition, there are file compression formats, such as ARC and ZIP. Data compression is also widely used in backup utilities, spreadsheet applications, and database management systems. Certain types of data, such as bit-mapped graphics, can be compressed to a small fraction of their normal size.

o **Lossy compression:** This Refers to data compression techniques in which some amount of data is lost. *Lossy compression* technologies attempt to eliminate redundant or unnecessary information. Most video compression technologies, such as *MPEG*, use a lossy technique.

o **Lossless:** *Lossless* refers to data compression techniques in which no data is lost. The PKZIP compression technology is an example of *lossless compression*. For most types of data, *lossless compression* techniques can reduce the space needed by only about 50%. For greater compression, one must use a *lossy compression* technique. Note, however, that only certain types of data such as graphics, audio, and video that can tolerate *lossy compression*. A *lossless compression* technique is used when compressing data and programs.

o **Decoder:** It is a device or program that translates encoded data into its original format (e.g., it *decodes* the data). The term is often used in reference to *MPEG-2* video and sound data, which must be decoded before it is output. Most DVD players, for example, include a *decoder card* whose sole function is to decode *MPEG* data. It is also possible to decode MPEG data in software, but this requires a powerful microprocessor.

o **High-Definition Television (HDTV):** It is a new type of television that provides much better resolution than current televisions based on the *NTSC* standard. There are a number of competing *HDTV* standards, which is one reason that the new technology has not been widely implemented. All of the standards support a wider screen than *NTSC* and roughly twice the resolution. To pump this additional data through the narrow TV channels, images are digitized and then compressed before they are transmitted and then decompressed when they reach the TV.

o **Digital Satellite System (DSS):** It is a network of satellites that broadcast digital data. An example of a *DSS* is *DirecTV,* which broadcasts digital television signals. *DSS's* are expected to become more important as the TV and computer converge into a single medium for information and entertainment.

o **National Television Standards Committee (NTSC):** The *NTSC* is responsible for setting television and video standards in the United States (in Europe and the rest of the world, the dominant television standards are *PAL* and *SECAM*). The *NTSC* standard for television defines a composite video signal with a refresh rate of 60 half-frames (interlaced) per second. Each frame contains 525 lines (at 60Hz frequency) and can contain 16 million different colors. The *NTSC* standard is incompatible with most computer video standards, which generally use *RGB* video signals with an *FM* frequency for audio. However, special video adapters can be inserted into the computer that converts *NTSC* signals into computer video signals and vice versa. The new digital television standard being developed is called *HDTV (High-Definition Television)*.

o **Phase Alternating Line (PAL):** It is the dominant television standard in Europe. PAL delivers 625 lines (at 50Hz frequency) at 50 half-frames per second. Many video adapters that enable computer monitors to be used as television screens support both NTSC and PAL signals.

o **Sequential Couleur Avec Mémoire (SECAM):** *SECAM* is the television broadcast standard in France, the Middle East, and most of Eastern Europe, *SECAM* broadcasts 819 lines of resolution per second.

o **Quarter Common Intermediate Format (QCIF):** *QCIF* is a videoconferencing format that specifies data rates of 30 frames per second (*fps*), with each frame containing 144 lines and 176 pixels per line. This is one fourth (¼) the resolution of *Full CIF*.

o **Videoconferencing:** This is conducting a conference between two or more participants at different sites by using computer networks to transmit audio and video data. For example, a *point-to-point* (two-person) video conferencing system works much like a video telephone. Each participant has a video camera, microphone, and speakers mounted on his or her computer. As the two participants speak to one another, their voices are carried over the network and delivered to the other's speakers, and whatever images appear in front of the video camera appear in a window on the other participant's monitor. Multipoint videoconferencing allows three or more participants to sit in a virtual conference room and communicate as if they were sitting right next to each other. Until the mid 90s, the hardware costs made videoconferencing prohibitively expensive for most organizations, but that situation is changing rapidly. Many analysts believe that videoconferencing will be one of the fastest-growing segments of the computer industry in the latter half of the decade.

o **H.323:** It is a standard approved by the *International Telecommunication Union (ITU)* that defines how audiovisual conferencing data is transmitted across networks. In theory, H.323 should enable users to participate in the same conference even though they are using different videoconferencing applications. Although most videoconferencing vendors have announced that their products will conform to H.323, it has not been proven whether such adherence will actually result in interoperability.

o **H.324:** This is a suite of standards approved by the *International Telecommunications Union (ITU)* that defines videoconferencing over analog (POTS) telephone wires. One of the main components of H.324 is the *V.80* protocol that specifies how modems should handle streaming audio and video data.

o **International Telecommunication Union (ITU):** It is an intergovernmental organization through which public and private organizations develop telecommunications. The ITU was founded in 1865 and became a United Nations agency in 1947. It is responsible for adopting international treaties, regulations and standards governing telecommunications. The standardization functions were formerly performed by a group within the *ITU* called *CCITT*, but after a 1992 reorganization the *CCITT* no longer exists as a separate body.

- **MP3**

MP3 is a standard technology and format for compressing a sound sequence into a very small file (about one-twelfth the size of the original file) while preserving the original level of sound quality when it is played. MP3 files (identified with the file name suffix of ".mp3") are available for downloading from Web sites. Windows 98 has a player built into the Operating System or it can be downloaded from one of several popular MP3 sites. MP3 files are usually download-and-play files rather than streaming sound files.

MP3 is the file extension for *MPEG, Audio Layer 3. Layer 3* is one of three coding schemes (*Layer 1, Layer 2 and Layer 3*) for the compression of audio signals. Layer 3 uses perceptual audio coding and psychoacoustic compression to remove all superfluous information (specifically the redundant and irrelevant parts of a sound signal). It also adds a *Modified Discrete Cosine Transform (MDCT)* that implements a filter bank, increasing the frequency resolution 18 times higher than that of layer 2. The result in real terms is layer 3 shrinks the original sound data from a CD (with a *bitrate* of 1411.2 kilobits per one second of stereo music) by a factor of 12 (down to 112-128kbps) without sacrificing sound quality. *Bitrate* denotes the average number of bits that one second of audio data will consume. Because MP3 files are small, they can easily be transferred across the Internet.

- **3-D Audio**

The *3-D Audio (or 3-D sound)* is a technique for giving more depth to traditional stereo sound. Typically, *3-D Audio* is generated by placing a device in a room with stereo speakers. The device dynamically analyzes the sound coming from the speakers and sends feedback to the sound system so that it can readjust the sound to give the impression that the speakers are further apart. *3-D audio* devices are particularly popular for improving computer audio where the speakers tend to be small and close together. There are a number of *3-D audio* devices that attach to a computer's sound card:

o **Dolby Digital:** This is a standard for high-quality digital audio that is used for the sound portion of video stored in a digital format, especially videos stored on DVD-ROMs. Dolby Digital delivers 6 channels in the "5:1" configuration: left, right, and center screen channels, separate left and right sounds, and a subwoofer channel. This is also called *surround sound* or *3D sound.*

o **Musical Instrument Digital Interface (MIDI):** *MIDI* is a standard adopted by the electronic music industry for controlling devices, such as synthesizers and sound cards that emit music. At minimum, a *MIDI* representation of a sound includes values for the note's pitch, length, and volume. It can also include additional characteristics, such as attack and delay time. The *MIDI* standard is supported by most synthesizers, i.e., sounds created on one synthesizer can be played and manipulated on another synthesizer. Computers that have a MIDI interface can record sounds created by a synthesizer and then manipulate the data to re-produce new sounds. For example, the key of a composition can be changed with a single keystroke. Software programs are available for composing and editing music that conforms to the *MIDI* standard. They offer a variety of functions: for instance, when a tune is played on a keyboard connected to a computer, a music program can translate what is played into a written score. *MIDI* was developed originally to allow synthesizers and digital instruments to be connected to computers. Many synthesizers can be connected to the sound card allowing the PC to compose MIDI files using the synthesizer. *MIDI* does not tell the sound card exactly what frequencies of sound to reproduce; rather, it tells the sound card to play notes as if the card were a musical instrument. Since there is no actual sound encoded in a MIDI file, they are much smaller in size than digitized audio files. There are two different ways that *MIDI* files can be played by a sound card:

- **Frequency Modulation (FM) Synthesis:** With this technique, instrument sounds are created by oscillators, much the same way that a conventional keyboard synthesizer works. FM synthesis can produce sounds reminiscent of the "real thing", but the sound is noticeably artificial, much the way a synthesized piano, harp or saxophone sounds artificial.

- **Wavetable Synthesis:** This more advanced technique uses a ROM chip on the sound card that contains an actual recorded sample of the real instrument to be played. When the sound card needs to play different notes it modifies the frequency of the sample, producing a much more realistic sound. Most wavetable sound cards also come with wavetable RAM (users can load instrument samples from disk, allowing them to specifically tailor the instrument to complement the piece they are writing/playing). The downfall to this is that anyone who tries to listen to that *MIDI* file must have their sound card loaded with the same instrument samples that user had, for it to sound the way it was intended.

There are also different *MIDI* standards; common ones include:

- Adlib

- Roland GS

- Roland MT32

- Microsoft Sound System

- EMU8000 -- 32 voices, wavetable

- GM -- General MIDI, FM

- MPU-401

- *RealAudio*

RealAudio is the current standard for streaming audio data over the World Wide Web. *RealAudio* was developed by RealNetworks and supports FM-stereo-quality sound. To hear a

Web page that includes a RealAudio sound file, a RealAudio player or plug-in is needed. A streaming technology developed by RealNetworks for transmitting live video over the Internet. RealVideo uses a variety of data compression techniques and works with both normal IP connections as well as IP Multicast connections.

- *Digital Audio*

Most people know that real-world sounds are analog, they do not have discrete values, but rather are continuous in nature. When someone hears a whistle, for example, the sound is continuous. If the whistle is stronger, the sound increases from quieter to louder in a gradual, smooth way. Computers are digital: they work with discrete values. Computerized sound uses discrete approximations to represent the analog world of "real" audio. There are also advantages to storing information digitally.

- *Digital and Analog Information*

There are two ways to represent information. Information that is continuous, that is, any information can take on any of an infinite set of values, is identified as *analog*. For example, the temperature (73 degrees outside, it could really be 73.31253 degrees, or any value between 73.00000 and 73.99999), the speed of a car (35 or 35.15425 mph), all of these have a continuous range of values. Digital information is restricted to a finite set of values. For example, a traffic light is (normally) red, yellow or green; not "yellow-green" or orange. Computers use a form of digital information called *binary* information. Here, the information is restricted to only two values: one or zero. Computers use binary information for several reasons:

- *Simplicity:* It is the simplest, most compact and least ambiguous way to express information about something: for example, zero is OFF and one is ON could be used to represent the status of a regular light bulb.

- *Expandability:* It is easy to build on and expand: two binary values can be used together to represent the status of two light bulbs.

- *Clarity:* Errors are reduced when a value can only be one or zero; the computer knows there are no values in between. If a modem line has a 0.98 value, the computer interprets that as a 1, since 0.98 is not a valid value and no data will be lost as a result.

- *Speed:* Computers make millions of decisions a second, and these decisions are easier to make when the number of values is small.

Digital information is often represented only in binary form, but does not have to be. An example is the compact disk audio, where sound information is stored as digital samples. The advantage of digital sampling is that the information is the same every time it is read, so there is no "loss" in quality over time as found in conventional magnetic analog storage media. Since analog information and digital information are certainly not the same, one is converted into the other. The process used to do this conversion is called *sampling*. When an analog signal is sampled, an electronic device measures the level of the signal periodically, and then records it in digital form. Then, by playing back the samples at the same rate, an approximation of the original signal is produced. This is in fact what happens with compact disk audio, with "wave" files in games. Each sample represents the *amplitude* (loudness, in essence) of the signal at that moment in time.

- ### *Streaming*

Streaming is a technique for transferring data such that it can be processed as a steady and continuous stream. Streaming technologies are becoming increasingly important with the growth of the Internet because most users do not have fast enough access to download large multimedia files quickly. With streaming, the client browser or plug-in can start displaying the data before the entire file has been transmitted. For streaming to work, the client side receiving the data must be able to collect the data and send it as a steady stream to the application that is processing the data and converting it to sound or pictures. This means that if the streaming client receives the data more quickly than required, it needs to save the excess data in a buffer. If the data does not come quickly enough, however, the presentation of the data will not be smooth.

- ### *ActiveMovie*

ActiveMovie is a new multimedia streaming technology developed by *Microsoft*. *ActiveMovie* supports most multimedia formats, including *MPEG*. It enables users to view multimedia content distributed over the Internet or CD-ROM.

- ### *WAV*

Wav is the format for storing sound in files developed jointly by *Microsoft* and *IBM*. Support for WAV files was built into Windows 95 making it the current standard for sound on PCs. WAV sound files end with a "*.wav*" extension and can be played by nearly all Windows applications that support sound.

- ### *Multimedia Personal Computer (MPC)*

MPC is a software and hardware standard developed by a consortium of computer firms led by *Microsoft*. There are three MPC standards, called *MPC*, *MPC2*, and *MPC3*, respectively. Each specifies a minimum hardware configuration for running multimedia software. To run *MPC-2* software, a minimum requirement of an Intel 486SX microprocessor with a clock speed of 25 MHz, 4 MB (megabytes) of RAM, a VGA display, and a double-speed CD-ROM drive. *MPC3* specifies the following minimum configuration of Pentium processor (75 MHz), 8 MB RAM, 540 MB disk drive, 4X CD-Rom and MPEG support.

- ### *Real-Time Transport Protocol (RTP)*

RTP is an internet protocol for transmitting real-time data such as audio and video. *RTP* itself does not guarantee real-time delivery of data, but it does provide mechanisms for the sending and receiving applications to support streaming data. Typically, *RTP* runs on top of the *UDP* protocol, although the specification is general enough to support other transport protocols.

- ### *Real Time Streaming Protocol (RTSP)*

RTSP is a proposed standard for controlling streaming data over the World Wide Web. *RTSP* was developed by Columbia University, Netscape and RealNetworks and has been submitted to the IETF for standardization. Like H.323, *RTSP* uses *RTP* to format packets of multimedia content. H.323 is designed for videoconferencing of moderately-sized groups, *RTSP* is designed to efficiently broadcast audio-visual data to large groups.

- ### *Audio Scrubbing*

Audio Scrubbing is the process of moving within an audio file or tape to locate a particular section. The term originally comes from the days of reel-to-reel players, when rocking a reel would give the impression of scrubbing tape across the head. Many audio scrub tools today allow the user to drag a cursor across the wave form to audition different sections of an audio file.

- ### *Audio (AU)*

AU is a common format for sound files on *UNIX* machines. It is also the standard audio file format for the Java programming language. *AU* files generally end with a *.au extension*. On PCs, two other popular sound formats are *WAV* and *MIDI*

- ### *Scanner*

A *scanner* (*Figure 13.2*) captures images from photographic prints, posters, magazine pages, and similar sources for computer editing and display. Scanners come in hand-held, feed-in, and flatbed types and for scanning black-and-white only or color. Very high resolution scanners are used for scanning for high-resolution printing, but lower resolution scanners are adequate for capturing images for computer display. Scanners usually come with software, such as *Adobe's Photoshop* product, that lets the user resizes and otherwise modifies a captured image. Scanners usually attach to a personal computer with a *Small Computer System Interface (SCSI)*. An application such as PhotoShop uses the *TWAIN* program to read in the image. Some major manufacturers of scanners include: *Epson, Hewlett-Packard, Microtek,* and *Relisys*.

Figure 13.2

The scanner is a device that can read text or illustrations printed on paper and translate the information into a form the computer can use. A scanner works by digitizing an image -- dividing it into a grid of boxes and representing each box with either a zero or a one, depending on whether the box is filled in. (For color and gray scaling, the same principle applies, but each box is then represented by up to 24 bits.) The resulting matrix of bits, called a bit map, can then be stored in a file, displayed on a screen, and manipulated by programs. Optical scanners do not distinguish text from illustrations; they represent all images as bit maps. Therefore, a text that has been scanned cannot directly be edited. To edit text read by an optical scanner, an *optical character recognition (OCR)* system is needed to translate the image into ASCII characters.

Carlos E. Hattab

Most optical scanners sold today come with OCR packages. Scanners differ from one another in the following respects:

- *scanning technology:* Most scanners use charge-coupled device (CCD) arrays, which consist of tightly packed rows of light receptors that can detect variations in light intensity and frequency. The quality of the CCD array is probably the single most important factor affecting the quality of the scanner. Industry-strength drum scanners use a different technology that relies on a photomultiplier tube (PMT), but this type of scanner is much more expensive than the more common CCD -based scanners.
- *resolution:* The denser the bit map, the higher the resolution. Typically, scanners support resolutions of from 72 to 600 dpi.
- *bit depth:* The number of bits used to represent each pixel. The greater the bit depth, the more colors or grayscales can be represented. For example, a 24-bit color scanner can represent 2 to the 24th power (16.7 million) colors. Note, however, that a large color range is useless if the CCD arrays are capable of detecting only a small number of distinct colors.
- *size and shape:* Some scanners are small hand-held devices that can be moved across the paper. These hand-held scanners are often called *half-page* scanners because they can only scan 2 to 5 inches at a time. Hand-held scanners are adequate for small pictures and photos, but they are difficult to use if it is needed to scan an entire page of text or graphics.

Larger scanners include machines into which can be fed sheets of paper. These are called *sheet-fed* scanners. *Sheet-fed* scanners are excellent for loose sheets of paper, but they are unable to handle bound documents. A second type of large scanner, called a *flatbed scanner*, is like a photocopy machine. It consists of a board on which books, magazines, and other documents are laid that can be scanned. Flatbed scanners are particularly effective for bound documents. *Overhead* scanners (also called *copyboard* scanners) look somewhat like overhead projectors. Documents are placed face-up on a scanning bed, and a small overhead tower moves across the page.

- *Optical Character Recognition (OCR)*

OCR refers to the branch of computer science that involves reading text from paper and translating the images into a form that the computer can manipulate (for example, into ASCII codes). An OCR system enables the user to take a book or a magazine article, feed it directly into an electronic computer file, and then edit the file using a word processor. All OCR systems include an optical scanner for reading text, and sophisticated software for analyzing images. Most OCR systems use a combination of hardware (specialized circuit boards) and software to recognize characters, although some inexpensive systems do it entirely through software. Advanced OCR systems can read text in large variety of fonts, but they still have difficulty with handwritten text. The potential of OCR systems is enormous because they enable users to harness the power of computers to access printed documents. OCR is already being used widely in the legal profession, where searches that once required hours or days can now be accomplished in a few seconds.

- *Technology Without An Important Name (TWAIN)*

TWAIN is a program that lets the user scan an image (using a scanner) directly into the application (such as PhotoShop) where the image can be viewed or edited. Without *TWAIN*, an

144

application that was open, would have to be closed, open a special application to receive the image, and then move the image to the application where it is wanted to work with it. The *TWAIN* driver runs between an application and the scanner hardware. *TWAIN* usually comes as part of the software package when a scanner is purchased. It is also integrated into PhotoShop and similar image manipulation programs. The software was developed by a work group from major scanner manufacturers and scanning software developers and is now an industry standard.

- ## *Digital Camera*

The *Digital Camera* (*Figure 13.3*) is a new technology that swept the PC users with videoconferencing capabilities. It is a camera that stores images digitally rather than recording them on film. The *Digital camera* has revolutionized the computers to another level of high-speed technology. In a typical videoconference session, a camera and a microphone capture the picture and sounds of the transmitter user, send analog signals to a video-capture adapter board. Typical digital camera processes between 5 and 15 *fps* compare to the TV's and movies' of 30 *fps*. The reason is to cut down on the amount of data that must be processed by the PC. In some video sessions, the picture may look choppy or jerky. That is because the PC is processing about half the number of frames per second that movies or TVs use. The frame rate is slower than the human eye is accustomed to seeing. Once the video-capture adapter board captures the frames, an *Analog-to-Digital Converter (ADC)* chip converts the analog signals (video and audio signals) into digital signals (pattern of 1's and 0's). Then a *compression/decompression (codec)* chip or software reduces the amount of data needed to recreate the video signals. The software writes the video to disk by interweaving the data for the picture with the audio in a file format called ".avi" (audio/video interleave). To replay the video, the compression and the combined video/audio data is either sent through a *codec* chip or processed by software. Either method restores areas that had been eliminated by compression. The combined audio and video elements of the signal are separated and both are sent to a *DAC*, which translate the Binary data into analog signals that are then transmitted to the PC's monitor and speakers. Instead of being recorded, the compressed video and audio signals may be sent over the telephone line. For faster video and audio playback, those signals may be sent over a network or an *Integrated Services Digital Network (ISDN)* line because the data would be transmitted in a digital form (rather than being converted to an analog signal as used by ordinary telephone line).

Figure 13.3

Once a picture has been taken, it can be downloaded to a computer system, and then manipulated with a graphics program and printed. Unlike film photographs, which have an almost infinite resolution, digital photos are limited by the amount of memory in the camera, the optical resolution of the digitizing mechanism, and, finally, by the resolution of the final output device. Even the best digital cameras connected to the best printers cannot produce film-quality photos. However, if the final output device is a laser printer, it doesn't really matter whether a digital photo or a real photo is taken and then scanned. In both cases, the image must eventually be reduced to the resolution of the printer. The big advantage of digital cameras is that making photos is both inexpensive and fast because there is no film processing. Interestingly, one of the biggest boosters of digital photography is Kodak, the largest producer of film. Kodak developed the *Kodak PhotoCD* format, which has become the de facto standard for storing digital photographs. Most digital cameras use *CCDs* to capture images, though some of the newer less expensive cameras use *CMOS* chips instead.

Digitization is performed by sampling at discrete intervals. To translate an image into a digital form, for example, optical scanner digitizes the image by translating it into bit map. It is also possible to digitize sound, video, and any type of movement. To digitize sound, for example, a device measures a sound wave's amplitude many times per second. These numeric values can then be recorded digitally.

- ### *Charge-Coupled Device (CCD)*

The *CCD* is an instrument whose semiconductors are connected so that the output of one serves as the input of the next. Digital cameras, video cameras, and optical scanners all use *CCD* arrays.

- ### *Digital Audio Tape (DAT)*

DAT is a standard medium and technology for the digital recording of audio on tape at a professional level of quality. A *DAT* drive is a digital tape recorder with rotating heads similar to those found in a video deck. Most *DAT* drives can record at sample rates of 44.1 KHz, the CD audio standard, and 48 KHz. *DAT* has become the standard archiving technology in professional and semi-professional recording environments for master recordings. Digital inputs and outputs on professional *DAT* decks allow the user to transfer recordings from the *DAT* tape to an audio workstation for precise editing. The compact size and low cost of the *DAT* medium makes it an excellent way to compile the recordings that are going to be used to create a CD master.

CHAPTER XIV
INPUT DEVICES

Personal Computers are general-purpose machines. They are designed to perform a variety of tasks. Diverse groups of people (businesspeople, teachers, students, lawyers, doctors, farmers, store managers, etc..) purchase computers for many reasons (computing, documents, database, graphics, etc..). They may need different peripherals and software. They may need various types of information services. Certain peripherals devices work especially well in a Personal Computer environment despite some limitations such as cost and ease of use. The devices must be easy to use and at a reasonable cost. The increased use of the PC has promoted the popularity of a variety of input and output devices, many of which have become essential for easy use of the PC.

- *Mouse*

One of the input devices that became a very important part of the Personal Computer is the *"Mouse"*. It is a device that controls the movement of the cursor or pointer on a display screen. Invented by Douglas Engelbart of Stanford Research Center in 1963, and pioneered by Xerox in the 1970s, the mouse is one of the great breakthroughs in computer ergonomics, because it frees the user to a large extent from using the keyboard. In particular, the mouse is important for graphical user interfaces because the PC user can simply point to options and objects and click a mouse button. Such applications are often called *point-and-click* programs. The mouse is also useful for graphics programs that allow the user to draw pictures by using the mouse like a pen, pencil, or paintbrush.

A mouse is a small object that can roll along a hard, flat surface. Its name is derived from its shape, which looks a bit like a mouse, its connecting wire that one can imagine to be the mouse's tail, and the fact that one must make it scurry along a surface. As the mouse is moved, the pointer on the display screen moves in the same direction. A mouse, typically, contains at least one button and sometimes as many as three, which have different functions depending on what program is running. Some new mouse devices also include a *scroll wheel* for scrolling through long documents.

The *mouse* (*Figure 14.0*) is a pointing device that when positioned by hand, translates that physical position into a cursor position on the computer monitor. It is a movable input device, normally by a hand or a finger, that controls the position of the computer *cursor* on the monitor.

A *cursor* is the position indicator on a computer display screen where a user can enter text. In an operating system with a graphical user interface (GUI), the cursor is also a visible and moving pointer that the user controls with a mouse, a touch pad, or a similar input device. The user uses the pointing cursor and special input buttons to establish where the position indicator cursor will be or to select a particular program to run or file to view. Typically, the pointing cursor is an arrow and the text entry position cursor is a blinking underscore or vertical bar. Most operating systems allow the user to choose another appearance for the cursor. In the Windows operating system, the pointing cursor turns into a pointing hand when it is pointed to a specific object and into an hourglass while the system is starting a requested program. For example, in a document, the "cursor" is a symbol on a computer's monitor that shows where the next typed character will appear.

Typically, on the bottom of the mouse, there is a small ball similar to a roller bearing. On the top, however, there are one, two or three buttons which are used to activate commands such as highlights, cut, past, etc.. When the mouse is moved (rolled across a flat surface), the cursor on the monitor moves accordingly. Using the mouse has eliminated a considerable amount of typing.

Figure 14.0

There are three basic types of mouse devices:

o **Mechanical:** This type of mouse has a rubber or metal ball on its underside that can roll in all directions. Mechanical sensors within the mouse detect the direction the ball is rolling and move the screen pointer accordingly.

o **Optomechanical:** This mouse is the same as a mechanical mouse, but uses optical sensors to detect motion of the ball.

o **Optical:** This mouse uses a laser to detect the mouse's movement. The mouse must be moved along a special mat with a grid so that the optical mechanism has a frame of reference. An optical mouse has no mechanical moving parts. It responds more quickly and precisely than mechanical and optomechanical mouse devices, but it is also more expensive.

A mouse connects to PCs in one of these ways:

o **Serial mouse:** It connects directly to an RS-232C serial port or a PS/2 port. This is the simplest type of connection.

o **PS/2 mouse:** It connects to a PS/2 port (*Figure 14.1*).

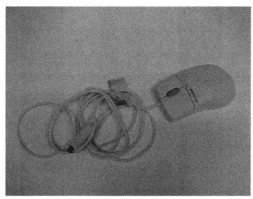

Figure 14.1

o **Cordless mouse:** It is not physically connected at all. Instead it relies on infrared or radio waves to communicate with the computer. A cordless mouse is more expensive than both serial and bus mouse devices, but it eliminates the cord.

o **Bus mouse:** It connects to the computer using an *expansion board* plugged into an *expansion slot (Figure 14.2).*

Figure 14.2

Although the mouse has become a familiar part of the personal computer, its design continues to evolve and there continue to be other approaches to pointing or positioning on a display. Notebook computers include built-in mouse devices that let the user controls the cursor by finger over a built-in *trackball*. Users of graphic design and CAD applications can use a stylus and a specially-sensitive pad to draw as well as move the cursor. Other display screen-positioning ideas include a video camera that tracks the user's eye movement and places the cursor accordingly.

- *Mousepad*

A *mousepad* is a small, portable surface that sometimes provides better traction for the ball on a computer mouse and, at the very least, provides a bounded area in which to move the mouse. On very smooth table or desktop surfaces, the mouse ball may tend to glide instead of roll unless pressure is continually applied. Some mousepads provide a slightly rougher surface so that it's easier to get the mouse to move the screen cursor. A mousepad usually has a plastic surface and a thin rubber or plastic cushion. Mousepads are sometimes given away at computer and Internet trade shows or with Internet or computer magazine subscriptions.

- **Trackball**

A *trackball* is a computer cursor control device used in many notebook and laptop computers. The trackball is usually located in front of the keyboard toward the user. Essentially, the trackball is an upside-down mouse that rotates in place within a socket. The user rolls the ball to direct the cursor to the desired place on the screen and can click one of two buttons (identical to mouse buttons) near the trackball to select desktop objects or position the cursor for text entry. IBM's ThinkPad series of notebook computers uses a *"pointing stick"*, called a *TrackPoint*, that is integrated into the middle of the keyboard keys.

- **Touch Screen**

A *Touch Screen* is a type of display screen that has a touch-sensitive transparent panel covering the screen. Instead of using a pointing device such as a mouse or light pen, a finger can be used to point directly to objects on the screen.

- **Touch pad**

The *Touch pad* was invented by George E. Gerpheide in 1988. It is a small, touch-sensitive pad used as a pointing device on some portable computers. By moving a finger or other object along the pad, the pointer can be moved on the display screen. The user clicks by tapping the pad. A *touch pad* is a device for pointing (controlling input positioning) on a computer display screen. It is an alternative to the mouse. Originally incorporated in laptop computers, touch pads are also being made for use with desktop computers. A touch pad works by sensing the user's finger movement and downward pressure. *Apple Computer* was the first to license and use the *touch pad* in its Powerbook laptops in 1994. The *touch pad* has since become the leading cursor controlling device in laptops. Many laptops use a *trackball*. *IBM ThinkPad* laptops use a "pointing stick" (called a *TrackPoint*) that is set into the keyboard.

The *touch pad* contains several layers of material. The top layer is the pad that is touched by the user. Beneath it are layers (separated by very thin insulation) containing horizontal and vertical rows of *electrodes* that form a grid. Beneath these layers is a circuit board to which the electrode layers are connected. The layers with electrodes are charged with a constant *alternating current (AC)*. As the finger approaches the electrode grid, the current is interrupted and the interruption is detected by the circuit board. The initial location where the finger touches the pad is registered so that subsequent finger movement will be related to that initial point. Some touch pads contain two special places where applied pressure corresponds to clicking a left or right mouse button. Other touch pads sense single or double taps of the finger at any point on the touch pad.

- **Graphics tablet**

A *Graphics tablet* is a flat, boardlike surface directly connected to the PC's monitor. The user draws on a tablet using a pencil-like device, and an image is transmitted to the display of the PC. A *graphics tablet* enables the user to employ colors, textures and patterns when creating an image. A typical application for *graphics tablet* is the *CAD/CAM* system.

- **Light pen**

A *Light pen* is an input device that utilizes a light-sensitive detector to select objects on a display screen. A light pen is similar to a mouse, except that with a light pen the user can move the pointer and select objects on the display screen by directly pointing to the objects with the pen.

- *Keyboard*

The *keyboard* (*Figure 14.3*) is, on most computers, the primary text input device. The mouse is also a primary input device but lacks the ability to easily transmit textual information. The keyboard contains certain standard function keys, such as the *Escape* key, *tab* and *cursor movement* keys, *shift* and *control* keys, and some customized keys.

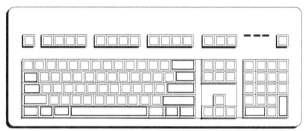

Figure 14.3

The computer's keyboard uses the same key arrangement as the mechanical and electronic typewriter keyboards that preceded the computer. The standard arrangement of alphabetic keys is known as the *Qwerty* keyboard, its name deriving from the arrangement of the five keys at the upper left of the three rows of alphabetic keys. This arrangement, invented for one of the earliest mechanical typewriters, dates back to the 1870s. Another well-known key arrangement is the *Dvorak* system, which was designed to be easier to learn and use. The *Dvorak* keyboard was designed with the most common consonants on one side of the middle or home row and the vowels on the other side so that typing tends to alternate key strokes back and forth between hands. Although the *Dvorak* keyboard has never been widely used, it has adherents. Because many keyboard users develop a cumulative trauma disorder, such as *carpal tunnel syndrome*, a number of ergonomic keyboards have been developed. Approaches include keyboards contoured to alleviate stress and foot-driven pedals for certain keys or keyboard functions.

- o **Ergonomic:** The ergonomic aspect of computers deals with their usability for humans. Ergonomics is the use of research in designing systems, programs, or devices that are easy to use for their intended purposes and contexts. The terms *human factors* and *usability* are related.

CHAPTER XV
MONITOR

In any computer system, the most common output device is the *Monitor*. The most important component in the monitor is the picture tube, also called a *Cathode Ray Tube* or *CRT*. The *CRT* is what makes the image that is seen on the screen, and its characteristics primarily determine the quality of the image viewed. A good monitor will always have a good *CRT*; no amount of fancy controls and other features can make up for a bad tube. In fact, the CRT defines the whole monitor so much that often the monitor is just called "the CRT". A *monitor* is a term for display screen of a computer. The term *monitor,* however, usually refers to the entire box, whereas display screen can mean just the screen. In addition, the term *monitor* often implies graphics capabilities. The cursor on the monitor is a special symbol that indicates the current screen position for the next entry or file display. The cursor can appear on the CRT in several forms of characters, depending on the program (an underline, an arrow, a blinking block, etc...). There are two types of monitors:

- **Desktop Monitor:** (also called *Cathode Ray Tube or CRT*), is a monitor (*Figure 15.0*) that displays the communications of the computer. A character or a graphics image is shown on the CRT that is sent by the computer through a video graphics card. It is the device that allows users to view information before sending it to the microprocessor for processing, or after the microprocessor has processed the information and sent the result back for users to view it. This information can be in either alphanumeric character or graphical form.

Figure 15.0

- **Liquid Crystal Display (LCD) Monitor:** A monitor that uses LCD technologies rather than the conventional CRT technologies used by most desktop monitors. This is the technology used for displays in notebook and other smaller computers. Like *Light-Emitting Diode (LED)* and Gas-plasma technologies, *LCD*s allow displays to be much thinner than cathode ray tube (CRT) technology. They consume much less power than *LED* and gas-display displays because they work on the principle of blocking light rather than emitting it. These are flat panel displays that were primarily use for portable computers (*Figure 15.1*). They are less bulky and require less power that the typical desktop monitors. Until recently, LCD panels were used exclusively on Notebook computers and other portable devices. In 1997, however, several manufacturers began offering full-size LCD monitors as alternatives to CRT monitors. The main advantage of LCD displays is that they take up less desk space and are lighter. An LCD is manufactured with either a *Passive Matrix* or an *Active Matrix* display grid.

Figure 15.1

LCD is a type of display used in digital watches and many portable computers. LCD displays utilize two sheets of polarizing material with a liquid crystal solution between them. An electric current passed through the liquid causes the crystals to align so that light cannot pass through them. Each crystal, therefore, is like a shutter, either allowing light to pass through or blocking the light. Monochrome LCD images usually appear as blue or dark gray images on top of a grayish-white background. Color LCD displays use two basic techniques for producing color: *Passive matrix* is the less expensive of the two technologies. The other technology, called *thin film transistor* (TFT) or *active-matrix*, produces color images that are as sharp as traditional CRT displays, but the technology is expensive. Recent passive-matrix displays using new CSTN and DSTN technologies produce sharp colors rivaling active-matrix displays. Most LCD screens used in notebook computers are backlit to make them easier to read.

- **Passive Matrix Display**: A common type of flat-panel display consisting of a grid of horizontal and vertical wires. At the intersection of each grid is an LCD element, which constitutes a single pixel, either letting light through or blocking it. A higher quality and more expensive type of display, called an *active-matrix display*, uses a transistor to control each pixel. In the mid-90s, it appeared that passive-matrix displays would eventually become extinct due to the higher quality of active-matrix displays. However, the high cost of producing active-matrix displays, and new technologies such as DSTN, CSTN and HPA that improve passive-matrix displays, have cause passive-matrix displays to make a surprising comeback.

- The **Dual Scan LCD** has a grid of conductors with pixels located at each intersection in the grid. A current is sent across two conductors on the grid to control the light for any pixel. Some *Passive Matrix* LCD's have dual scanning, meaning that they scan the grid twice with current in the same time that it took for one scan in the original technology. It is type of passive-matrix LCD display that provides faster refresh rates those conventional passive-matrix displays by dividing the screen into two sections that are refreshed simultaneously. Dual-scan displays are not as sharp or bright as active-matrix displays, but they consume less power.

- The **Active Matrix LCD** is also known as a *Thin Film Transistor* or *TFT* display. An active matrix has a transistor located at each pixel intersection, requiring less current to control the luminance of a pixel. Having a transistor at each pixel means that the current

153

that triggers pixel illumination can be smaller and therefore can be switched on and off more quickly. For this reason, the current in an *Active Matrix* display can be switched ON and OFF more frequently, improving the screen refresh time. The display provides a more responsive image at a wider range of viewing angle than a *Dual Scan* display.

o The **Plasma** display is a type of flat-panel display that works by sandwiching a neon/xenon gas mixture between two sealed glass plates with parallel electrodes deposited on their surfaces. The plates are sealed so that the electrodes form right angles, creating pixels. When a voltage pulse passes between two electrodes, the gas breaks down and produces weakly ionized plasma, which emits UV radiation. The UV radiation activates color phosphors and visible light is emitted from each pixel. Today, Plasma displays are becoming more and more popular. Compared to conventional CRT displays, plasma displays are about one-tenth the thickness--around 4", and one-sixth the weight--less than 67 pounds for a 40" display. They use over 16 million colors and have a 160 degree-viewing angle. Companies such as Panasonic, Fujitsu, and Pioneer manufacture plasma displays.

o The **Gas-Plasma Display** is a type of thin display screen, used in some older portable computers. A *Gas-plasma* display works by sandwiching neon gas between two plates. Each plate is coated with a conductive print. The print on one plate contains vertical conductive lines and the other plate has horizontal lines. Together, the two plates form a grid. When electric current is passed through a horizontal and vertical line, the gas at the intersection glows, creating a point of light, or *pixel*. Images on gas-plasma displays generally appear as orange objects on top of a black background. Although gas-plasma displays produce very sharp monochrome images, they require much more power than the more common LCD displays.

o The **Electroluminescent Display (ELD)** is a technology used to produce a very thin display screen used in some portable computers. An *ELD* works by sandwiching a thin film of phosphorescent substance between two plates. One plate is coated with vertical wires and the other with horizontal wires, forming a grid. When an electrical current is passed through a horizontal and vertical wire, the phosphorescent film at the intersection glows, creating a point of light, or *pixel*.

o **Color Super-Twist Nematic (CSTN)** is an LCD technology developed by Sharp Electronics Corporation. Unlike TFT, CSTN is based on a passive matrix, which is less expensive to produce. The original CSTN displays developed in the early 90's suffered from slow response times and ghosting. Recent advances in the technology, however, have made CSTN a viable alternative to active-matrix displays. New CSTN displays offer 100ms response times, a 140-degree viewing angle, and high-quality color rivaling TFT displays - all at about half the cost. A newer passive-matrix technology called *High-Performance Addressing (HPA)* offers even better response times and contrast than CSTN.

o **Double-layer Super-Twist Nematic (DSTN)** is a passive-matrix LCD technology that uses two display layers to counteract the color shifting that occurs with conventional supertwist displays.

o **High-Performance Addressing (HPA)** is a passive-matrix display technology that provides better response rates and contrast than conventional LCD displays. Although HPA displays aren't quite as crisp or fast as *active-matrix (TFT)* displays, they're considerably less expensive to produce. Consequently, HPA is being used by a number of computer manufacturers for their low-end notebook computers.

- o **Supertwist** is a technique for improving LCD display screens by twisting light rays. In addition to normal *supertwist* displays, there also exist *double supertwist* and *triple supertwist* displays. In general, the more twists, the higher the contrast. Supertwist displays are also known as *supertwist nematic (STN)* displays.
- o **Backlighting** is a technique used to make flat-panel displays easier to read. A backlit display is illuminated so that the foreground appears sharper in contrast with the background.

There are many ways to classify monitors. The most basic is in terms of color capabilities, which separates monitors into three classes:

- **Monochrome:** Monochrome monitors actually display two colors only, one for the background (color is usually black) and one for the foreground (colors are white, green, or amber). These monitors were very popular during the 1970's computer era because they were inexpensive and required no additional video circuitry inside the computer. They are another type of composite monitors, single video signal.
- **Gray-Scale:** A gray-scale monitor is a special type of monochrome monitor capable of displaying different shades of gray.
- **Color:** A display monitor is capable of displaying many colors. Color monitors can display anywhere from 4 to over 10 million different colors. Color monitors implement the *RGB* color model by using three different phosphors that appear **Red**, **Green**, and **Blue** when activated. By placing the phosphors directly next to each other, and activating them with different intensities, color monitors can create an unlimited number of colors. In reality, the video adapter controls the number of colors that any monitor can display. A *True Color* is the specification of the color of a *Pixel* on a display screen using a 24-bit value, which allows the possibility of up to 16,777,216 (over 16 million) possible colors. Many displays support only an 8-bit color value, allowing up to 256 possible colors. The number of bits used to define a pixel's color shade is its *bit-depth*. True color is sometimes known as *24-bit color*. Some new color display systems offer a 32-bit color mode. The extra byte, called the *alpha channel*, is used for control and special effects information. Color monitors based on CRT technology employ three different techniques to merge phosphor triplets into pixels:
  - o **Dot-trio shadow masks** place a thin sheet of perforated metal in front of the screen. Since electrons can pass only through the holes in the sheet, each hole represents a single pixel.
  - o **Aperture-grille CRTs** place a grid of wires between the screen and the electron guns.
  - o **Slot-mask CRTs** uses a shadow mask but the holes are long and thin. It's sort of a cross between the dot-trio shadow mask and aperture-grill techniques.

There are two types of color monitors:

- o **Composite Color:** Composite color monitors display a composite of colors received in a single video signal and are slightly more expensive than *monochrome* monitors. The display quality of text, however, is not as crisp as the monochrome, and the graphics quality is not as sharp as the RGB monitors.
- o **RGB (Red-Green-Blue):** These monitors typically receive three separate color signals, one for each of the three main colors: *Red*, *Green* and *Blue*. This differs from

155

color televisions, for example, which use composite video signals, in which all the colors are mixed together. All color computer monitors are RGB monitors. Commonly used for high-quality graphics displays, they display sharper images than the *Composite Color* monitors. This type of monitor normally requires a video adapter card inside the computer. An RGB monitor consists of a vacuum tube with three electron guns -- one each for red, green, and blue -- at one end and the screen at the other end. The three electron guns fire electrons at the screen, which contains a phosphorous coating. When the electron beams excite the phosphors, they glow. Depending on which beam excites them, they glow red, green, or blue. Ideally, the three beams should converge for each point on the screen so that each pixel is a combination of the three colors.

After this classification, the most important aspect of a monitor is its screen size. Like televisions, screen sizes are measured in diagonal inches, the distance from one corner to the opposite corner diagonally. Size ranges from a few inches to over 40 inches. An example size for a small VGA monitor is 14 inches. Monitors that are 16 or more inches diagonally are often called *full-page* monitors. In addition to their size, monitors can be either *Portrait* (height greater than width) or *Landscape* (width greater than height). Larger landscape monitors can display two full pages, side by side. The screen size is sometimes misleading because there is always an area around the edge of the screen that can't be used. Therefore, monitor manufacturers must now also state the viewable area -- that is, the area of screen that is actually used.

- **Picture Element or Pixel**

The image that is displayed on the screen is composed of thousands (or millions) of small dots; these are called *pixels*. A *Pixel or Pel* (from "*Picture Element*") is the basic unit of programmable color on a computer display or in a computer image. A *Pixel* is the smallest area, the shape of a dot, of the monitor's screen. This pixel can be turned ON or OFF. Changing the color of pixels creates the image on the computer's monitor. The higher the number of pixels, the better is the graphical image on the screen. Each one can be set to a different color and intensity (brightness). The number of pixels that can be displayed on the screen is referred to as the *resolution* of the image; this is normally displayed as a pair of numbers, such as 640x480. The first is the number of pixels that can be displayed horizontally on the screen, and the second how many can be displayed vertically. The higher the resolution, the more pixels that can be displayed and therefore the more that can be shown on the monitor at once, however, pixels are smaller at high resolution and detail can be hard to make out on smaller screens. Resolutions generally fall into predefined standard sets; only a few different resolutions are used by most PCs.

Resolution	Number of Pixels
640 x 480	307,200
800 x 600	480,000
1024 x 768	786,432
1280 x 1024	1,310,720

A monitor that the screen size is 17" (diagonal) can fit 307,200 pixels at 640 x 480 resolution or 1,310,720 pixels at 1280 x 1024 resolution. The physical size of a pixel depends on the resolution for the display screen. The physical size of a pixel at maximum resolution is equal the physical size of the dot pitch of the display. The lower the resolution, the larger the physical size of a pixel will be. The specific color that a pixel describes is a blend of three components of the color spectrum - *Red, Green,* and *Blue*. Up to three bytes of data are allocated for specifying a pixel's color, one byte for each color. A *true color* or *24-bit color* system uses all three bytes. However, most color display systems use only eight-bits (which provides up to 256 different colors).

The *aspect ratio* of the image is the ratio of the number of X pixels to the number of Y pixels. The standard aspect ratio for PCs is 4:3, but some resolutions use a ratio of 5:4. Monitors are calibrated to this standard so that if a circle is drawn, it will appear to be a circle and not an ellipse. Displaying an image that uses an aspect ratio of 5:4 will cause the image to appear somewhat distorted. The only mainstream resolution that currently uses 5:4 is the high-resolution 1280x1024. There is some confusion regarding the use of the term "resolution", since it can technically mean different things.

First, the resolution of the image seen is a function of what the video card outputs and what the monitor is capable of displaying; to see a high resolution image such as 1280x1024 requires both a video card capable of producing an image this large and a monitor capable of displaying it.

Second, since each pixel is displayed on the monitor as a set of three individual dots (red, green and blue), some people use the term "resolution" to refer to the resolution of the monitor, and the term "pixel addressability" to refer to the number of discrete elements the video card produces. In practical terms most people use resolution to refer to the video image. *Table 15.0* lists the most common resolutions used on PCs and the number of pixels each uses:

Resolution	Number of Pixels	Aspect Ratio
320x200	64,000	8:05
640x480	307,200	4:03
800x600	480,000	4:03
1024x768	786,432	4:03
1280x1024	1,310,720	5:04
1600x1200	1,920,000	4:03

Table 15.0

- *Pixel Color and Intensity*

Each pixel of the screen image is displayed on a monitor using a combination of three different color signals: red, green and blue. This is similar (but by no means identical) to how images are displayed on a television set. Each pixel's appearance is controlled by the intensity of these three beams of light. When all are set to the highest level the result is white; when all are set to zero the pixel is black, etc.

Carlos E. Hattab

- *Color Depth and the Color Palette*

The amount of information that is stored about a pixel determines its *color depth*, which controls how precisely the pixel's color can be specified. This is also sometimes called the *bit depth*, because the precision of color depth is specified in bits. The more bits that are used per pixel, the finer the color detail of the image. However, increased color depths also require significantly more memory for storage of the image, and also more data for the video card to process, which reduces the possible maximum refresh rate. *Table 15.1* shows the color depths used in PCs:

Color Depth	Number of Displayed Colors	Bytes of Storage Per Pixel	Common Name for Color Depth
4-Bit	16	0.5	Standard VGA
8-Bit	256	1	256-Color Mode
16-Bit	65,536	2	High Color
24-Bit	16,777,216	3	True Color

Table 15.1

True color is because three bytes of information are used, one for each of the red, blue and green signals that make up each pixel. Since a byte has 256 different values this means that each color can have 256 different intensities, allowing over 16 million different color possibilities. This allows for a very realistic representation of the color of images, with no compromises necessary and no restrictions on the number of colors an image can contain. In fact, 16 million colors are more than the human eye can discern. *True color* is popular for high-quality photo editing and graphical design. Some video cards actually have to use 32 bits of memory for each pixel when operating in true color, due to how they use the video memory

High color uses two bytes of information to store the intensity values for the three colors. Breaking the 16 bits into 5 bits for blue, 5 bits for red and 6 bits for green does it. This means 32 (or 2^5) different intensities for blue, 32 (or 2^5) for red, and 64 (or 2^6) for green. This reduced color precision results in a slight loss of visible image quality, but it is actually very slight because it is difficult to see the difference between *true color* and *high color* images. For this reason *high color* is often used instead of *true color*, it requires 33% to 50% less video memory, and it is also faster.

In *256-color mode* the PC has only 8 bits to use; this means 2 bits for blue and 3 for each of green and red, which results in choosing between only 4 or 8 different values for each color or block color, so a different approach is taken instead: the use of a *palette*. A palette is created containing 256 different colors. Each one is defined using the standard 3-byte color definition that is used in true color: 256 possible intensities for each of red, blue and green. Then, each pixel is allowed to choose one of the 256 colors in the palette, which can be considered a "color number" of sorts. The full range of color can be used in each image, but each image can only use 256 of the available 16 million different colors. When each pixel is displayed, the video card looks up the real red, green and blue values in the palette based on the "color number" the pixel is assigned. The palette is an excellent compromise: it allows only 8 bits to be used to specify each color in an image, but allows the creator of the image to decide what the 256 colors in the

158

image should be. Since virtually no images contain an even distribution of colors, this allows for more precision in an image by using more colors than would be possible by assigning each pixel a 2-bit value for blue and 3-bit values each for green and red. For example, an image of the sky with clouds (like the Windows 95 standard background) would have many different shades of blue, white and gray, and virtually no reds, greens, yellows and the like. 256-color is the standard for much of computing, mainly because the higher-precision color modes require more resources (especially video memory) and aren't supported by many PCs. Despite the ability to "hand pick" the 256 colors, this mode produces noticeably worse image quality than high color; most people *can* tell the difference between high color and 256-color mode.

- ### *Volume Pixel or Voxel*

A *Voxel* (from *"Volume Pixel"*) is a unit of graphic information that defines a point in three-dimensional space. Since a *Pixel* defines a point in two-dimensional space with its x and y coordinates. In a three-dimensional or *3-D* space, a z coordinate is needed. Each of the coordinates is defined in terms of its position, color, and density. Think of a cube where any point on an outer side is expressed with an x, y coordinate and the z coordinate defines a location into the cube from that side, its density, and its color. With this information and 3-D rendering software, a two-dimensional view from various angles of an image can be obtained and viewed at the computer. Medical practitioners and researchers are now using images defined by *Voxel* and 3-D software to view *X-rays*, *Cathode Tube Scans*, and *Magnetic Resonance Imaging* (*MRI*) scans from different angles, effectively to see the inside of the body from outside. Geologists can create 3-D views of earth profiles based on sound echoes. Engineers can view complex machinery and material structures to look for weaknesses.

The x and y coordinates are respectively the horizontal and the vertical addresses of any pixel or addressable point on a computer display screen. The x coordinate is a given number of pixels along the horizontal axis of a display starting from the pixel (*pixel 0*) on the extreme left of the screen. The y coordinate is a given number of pixels along the vertical axis of a display starting from the pixel (*pixel 0*) at the top of the screen. Together, the x and y coordinates locate any specific pixel location on the screen. x and y coordinates can also be specified as values relative to any starting point on the screen or any subset of the screen such as an image. On the Web, each click-able area of an image map is specified as a pair of x and y coordinates relative to the upper left-hand corner of the image.

A z coordinate is the third-dimensional coordinate in a volume *Pixel* or *Voxel*. Together with x and y coordinates, the z coordinate defines a location in a three-dimensional space.

- ### *Resolution*

The *Resolution* of a monitor indicates how densely packed the pixels are. *Resolution* is the number of *Pixels* contained on a display monitor, expressed in terms of the number of pixels on the horizontal axis and the number on the vertical axis. The sharpness of the image on a display depends on the resolution and the size of the monitor. The same pixel resolution will be sharper on a smaller monitor and gradually lose sharpness on larger monitors because the same number of pixels are being spread out over a larger number of inches In general, the more pixels (often expressed in dots per inch or dpi), the sharper is the image. Screen image sharpness is sometimes expressed as *dot per inch* (or *dpi*), the term *dot* means pixel, not dot as in *dot pitch. Both the*

159

Carlos E. Hattab

physical screen size and the resolution setting determine dot per inch. A given image will have less resolution - fewer dots per inch - on a larger screen as the same data is spread out over a larger physical area. Or, on the same size screen, the image will have less resolution if the resolution setting is made larger - resetting from 800 by 600 pixels per horizontal and vertical line to 640 by 480 means fewer dots per inch on the screen and an image that is less sharp. On the other hand, individual image elements such as text will be larger in size. Most modern monitors can display 1024 by 768 pixels, the SVGA standard. Some high-end models can display 1280 by 1024, or even 1600 by 1200. A given computer display system will have a maximum resolution that depends on its physical ability to focus light (in which case the physical dot size – the dot pitch - matches the pixel size) and usually several lesser resolutions. For example, a display system that supports a maximum resolution of "1280 by 1023" pixels may also support "1024 by 768", "800 by 600", and "640 by 480" resolutions. Note that on a given size monitor, the maximum resolution may offer a sharper image but be spread across a space too small to read well. Display resolution is not measured in dots per inch as it usually is with printers. However, the resolution and the physical monitor size together determine the pixels per inch. Typically, PC monitors have somewhere between 50 and 100 pixels per inch. For example, a 15-inch *VGA* monitor has a resolution of 640 pixels along a 12-inch horizontal line or about 53 pixels per inch. A smaller *VGA* display would have more pixels per inch.

Video RAM Required for Different Resolutions

Resolution	256 colors (8-bit)	65,000 colors (16-bit)	16.7 million colors (24-bit, true color)
640x480	512K	1 MB	1 MB
800x600	512K	1 MB	2 MB
1,024x768	1 MB	2 MB	4 MB
1,152x1,024	2 MB	2 MB	4 MB
1,280x1,024	2 MB	4 MB	4 MB
1,600x1,200	2 MB	4 MB	6 MB

Table 15.2

- *Analog and Digital Monitors*
  Another common way of classifying monitors is in terms of the type of signal they accept: *Analog* or *Digital*. Nearly all monitors accept analog signals, which is required by the VGA, SVGA, 8514/A and other high-resolution color standards. Today, all monitors are called *Analog* monitors, while older monitors are often called *Digital*. The electronic circuitry that controls the monitor is Digital. The *Color Signals* that are received from the video card are also Digital signals except the circuit that generates the video signals is Analog, called the *RAMDAC*. *Random Access Memory Digital-to-Analog Converter or RAMDAC,* is a single chip on video adapter cards. The *RAMDAC's* role is to convert digitally encoded images into analog signals that can be displayed by a monitor. A *RAMDAC* actually consists of four different components - SRAM to store the color map and three *Digital-to-Analog Converters* (*DACs*), one for each of the monitor's red, green, and blue electron guns (*Table 15.2*). Originally, monitors used digital

160

color signals, meaning each color had only a certain pre-set number of color levels that were supported. This was the case for *CGA* and *EGA* video cards and the monitors that work with them. Digital video signals are also sometimes called *"TTL"* for *"Transistor-to-Transistor Logic"*. Starting with *IBM's VGA* standard, the switch was made to analog color to allow for more possible shades of each of the three primary colors. An analog signal can have any of a continuous range of values, so the number of different color levels is in theory unlimited. In practice, standard analog color normally uses a range of 256 different color values for each color, yielding a total of 16.7 million different colors. All modern monitors and video cards use analog signaling. During the transition phase of the late 1980's, when there were a lot of *VGA* cards in use but also many older adapters, some companies put out monitors that can respond to both analog and digital signals. These usually use a toggle switch on the back of the case to select which mode they are working in. Most of these are quite old and obsolete at this point.

Note: Calling a monitor "Analog" or "Digital" refers to the type of color signals it uses. Since analog monitors can also have analog or digital *controls*, some companies confusingly call their monitors "Analog" or "Digital" based on this. However, this is inaccurate; in almost every case the monitor itself is analog either way.

- *Quality Factors*

A few monitors are *fixed frequency*, which means that they accept input signal at only one frequency. Most monitors, however, are multi-scanning, which means that they automatically adjust themselves to the frequency of the signals being sent to it. This means that they can display images at different resolutions, depending on the data being sent to them by the video adapters.

Other factors that determine a monitor's quality include the following:

- *Bandwidth:* The range of signal frequencies the monitor can handle. This determines how many Data signals it can process and therefore how fast it can refresh at higher resolutions.
- *Refresh Rate:* The *refresh rate* is the rate of how many times per second the screen is *refreshed* or *redrawn*. A higher rate translates to a more steady and "flicker free" display, but settings it too high may damage the monitor. *Table 15.3* shows typical settings:

Resolution setting	Refresh rate (Hz)	Horizontal scanning frequency (KHz)
640 x 480	72	37.8
800 x 600	72	48
1024 x 768	70	56
1280 x 1024	70	75

Table 15.3

- *Interlaced or Non-Interlaced:* Interlacing is a technique that enables a monitor to have more resolution, but it reduces the monitor's reaction speed.
  1) *Interlaced:* It is a method of producing the graphical image by overlapping two different image frames. The first is formed by scanning all the odd horizontal lines, while the second frame will be all the even horizontal lines.

161

This method usually requires half the refresh rate that is normally required to produce the non-interlaced mode. Interlaced monitors sometimes might cause flicker on the display as the monitor is used more and more.

2) ***Non-Interlaced:*** It is the opposite of the Interlaced monitor. A graphical image is formed by sequential scanning of all the horizontal lines. Non-Interlaced monitor usually requires twice the refresh rate of an Interlaced monitor but it is normally "flicker free" at the scanning rate of 72 Hz or higher.

o ***Dot Pitch, DPI or phosphor pitch:*** Most monitors are advertised with a dot pitch specification, usually from 0.25 to 0.40 dpi. *Dot Pitch* is the amount of space between each pixel. The smaller the dot pitch, the sharper is the image. It is a measurement that indicates the diagonal distance between like-colored phosphor dots on a display screen. Measured in millimeters, the dot pitch is one of the principal characteristics that determine the quality of display monitors. The lower the number, the crisper is the image. The dot pitch of color monitors for Personal Computers ranges from about 0.15 millimeters (mm) to 0.30 mm. The dot pitch specification for a display monitor specifies how sharp the displayed image can be. The dot pitch is measured in millimeters (mm) and a smaller number means a sharper image. In desktop monitors, common dot pitches are 0.31mm, 0.28mm, 0.27mm, 0.26mm, and 0.25mm. Small Monitors may have a 0.28mm dpi or finer. Some large monitors, for presentation use, may have a larger dot pitch (0.48mm, for example). Think of the dot specified by the dot pitch as the smallest physical visual component on the display. A *Pixel* is the smallest programmable visual element and maps to the dot if the display is set to its highest resolution. When set to lower resolutions, a pixel encompasses multiple dots. Technically, in a CRT display with a *shadow mask*, the dot pitch is the distance between the holes in the shadow mask, measured in millimeters (mm). The *shadow mask* is a metal screen filled with holes through which the three electron beams pass that focus to a single point on the tube's phosphor surface. The dot pitch of the monitor indicates how fine the dots are that make up the picture. The smaller the dot pitch, the sharper and detailed the image, all else being equal. Of course, all else is rarely equal, but the dot pitch is still one useful metric of the quality of a monitor.

o ***Shadow Mask and Aperture Grill:*** The monitor is made up of millions of tiny red, green and blue phosphor dots that glow when struck by the electron beam that travels across the screen to create the visible image. The dots are extremely small and the beam is traveling very quickly in order to cover the screen fast enough to allow for a smooth and stable image without flicker. To create a precise and crisp picture, it is necessary to make sure that the electron beam for each color strikes only the correct dots intended for use for that color. The normal way that this is done is by using a fine metal mesh called a *shadow mask*. The shadow mask is designed to the same shape as the surface of the CRT, and the electron beams shine through the mask. By carefully positioning the mask, the beams only strike the correct dots. The idea is similar to one way a sign can be made--a piece of paper is cut out with the shape of the letters and then laid on top of another surface. Then the paint is applied through the holes in the paper; the paper itself prevents the unwanted areas from being colored. An alternative way to accomplish the same task is taken by some CRTs. Instead of using a shadow mask they use what is called an *aperture grill*. Instead of a metal mesh, this type of tube uses many hundreds of fine metal strips that run vertically from the top of the screen surface to the bottom. These strips perform the same function as the shadow mask--they force the electron beam to illuminate only the correct parts of the screen. In CRTs that use an *aperture grill* (a slotted form of mask), such as Sony's Trinitron flat-screen technology, the dot pitch is the difference between adjacent slots that pass through an electron beam of the same color. The most common type of tube using this design is Sony's popular *Trinitron*, which is used in many brands of monitors. Compared to a shadow mask design, aperture grill CRTs have some advantages and one significant disadvantage. One advantage is that they allow more of the electron beam to pass through to the phosphor; this results in what many consider to be a brighter overall picture. Some also say that the picture on this type of monitor is sharper. Finally, because the strips are run straight from the top of the monitor to the bottom, this type of tube is flat vertically.

- o **Mask Pitch:** In color monitors, the distance between holes in the shadow mask. The *Mask Pitch* is essentially the same as the *Dot Pitch,* but it is measured on the mask rather than on the screen. The mask pitch is generally about 0.30 mm. The smaller the mask pitch, the sharper is the image.

- o **Convergence:** The clarity and sharpness of each pixel.

- **Quality Characteristics**

The selection of a monitor involves an image quality factors. *"A picture is worth a thousand words"* or *Pixels* in this case. When retailers advertise monitors for sale, they often include only a few specifications: the model number, size, and sometimes the viewable size and dot pitch. A monitor is one component that cannot be reliably purchased only based on the numbers, its consumer desirability is based on how it looks and this is not something that can be easily translated to figures, or even pictures. Buying a monitor is a subjective decision made by each person. It is best to purchase a monitor after seeing it in person.

Some of the other factors that influence the quality of the image produced by the monitor, many of these require personal testing by bringing up text or graphics on the screen to see what they look like, are:

- o **Sharpness / Focus:** One of the most important quality factors is how sharp images are. Virtually every decent monitor will produce a sharp image at lower resolution and in the center of the screen. Better ones will also produce sharp images at higher resolution and in the corners. If graphics or text is shown in the corners and the focus is noticeably worse than similar images at the center, the monitor is probably of lesser quality. If the monitor has a focus control however, make sure it is properly adjusted. Showroom floor monitors at retail stores are sometimes mercilessly tinkered with.

- o **Maximum Brightness:** Some monitors are limited in the maximum brightness level they can be set to. This can sometimes be a problem when using the monitor in a bright room. Also, the overall brightness level of the CRT will tend to decrease over time, so starting out with a bright monitor is better than a dim one.

- o **Straightness:** Vertical lines should be vertical, and horizontal lines horizontal. This isn't always the case, especially near the edges of the screen, where lines may bow inward or outward. Some monitors have a "pincushion" control that can be used to correct for this, but not all.

o **True Aspect Ratio:** Most monitors use a 4:3 aspect ratio, matching the aspect ratio of most popular screen resolutions. This is done to ensure that objects have the proper proportion of height to width. If a circle is drawn using a graphics program such as Paint, it should appear as a circle, and not as an ellipse.

o **Glare:** Monitors employ different chemical treatments and other techniques to reduce or eliminate surface glare. Glare, especially from overhead fluorescent lighting, can cause eyestrain and fatigue. Monitors that use a Trinitron CRT are vertically flat, which can reduce glare.

o **Distortion:** Many cheaper monitors have noticeable image distortion at their edges.

o **Color Purity:** A full screen of Red, Green or Blue should appear as Red, Green or Blue.

o **Monochrome Purity:** Some monitors do a poor job of displaying black text on a white background, or vice-versa: they can show color at the edges of letters.

- *Electrical Characteristics*

A quality monitor requires that its electrical characteristics be defined and measured as well. Some of the characteristics are:

o **Degaussing and Magnetization:** Degauss is to remove magnetism from a device. The term is usually used in reference to color monitor and other display devices that use a CRT. These devices aim electrons onto the display screen by creating magnetic fields inside the CRT. External magnetic forces -- such as the earth's natural magnetism or a magnet placed close to the monitor -- can magnetize the shadow mask, causing distorted images and colors. Since it may be impossible to remove the external magnetic force, degaussing works by re-aligning the magnetic fields inside the CRT to compensate for the external magnetism. Degauss can also remove all data from the magnetic media, such as floppy disks. The image quality of the monitor, especially the color, can be adversely affected if the internal components become magnetized. This can happen from exposure to a magnetic field (example: putting a magnet near the monitor, or through physical shock to the CRT by shaking it, and sometimes even by something as simple as changing its orientation). Magnetization manifests itself through splotches of color on the screen, especially in the corners. The process of eliminating magnetization on a CRT is called *degaussing* (the unit of measure of magnetic inductive force is the *Gauss*, named for mathematician Karl Friedrich Gauss). Most modern CRTs today include built-in degaussing circuits. Some have a manual switch to activate the circuit, some do it automatically at startup and others offer both as an option. The degaussing circuit uses a coil of wire to neutralize magnetic fields within the CRT. Manual degauss button can be used to degauss the monitor if it has magnetization (color) problems. Too many Degaussing in a row could damage the Degauss circuit or the monitor itself. When the degauss is engaged a buzzing sound is heard and the screen image will appear to vibrate for a few seconds; then a click is sounded, the buzzing will stop and the screen will return to normal. Some monitors that Degauss automatically at startup (power is turned ON), is a measure "preventive maintenance" function. This is accomplished at the power ON by listening to a buzzing then a click sounds in the first five seconds, this the degaussing circuit in action. Sometimes magnetization will not go away after a degaussing. If a monitor automatically degausses each time it is turned ON, sometimes the best course of action is to ignore the color impurity and wait for a week or two to see if over time, the multiple Degaussing during power-ON cycles will eliminate the problem. The other option is to use a Manual Degausser, which is a special demagnetizing device that is moved over the surface of the CRT to eliminate magnetic fields.

Warning: Manual Degaussers produce magnetic fields that can be damaging to data stored on magnetic media. Keep them well away from floppy disks and hard disks. It is also a good idea to just generally keep magnetic media away from the CRT. The opposite of pincushion distortion is *barrel distortion,* in which horizontal and vertical lines bend outwards toward the edge of the display. A third type of distortion, called *trapezoid distortion,* occurs when vertical lines are straight but not parallel with each other. Most monitors have *pincushion/barrel* controls that let the user correct these distortions, and many monitors also include a trapezoid distortion control.

o   ***Power and Safety:*** Due to the nature of how CRTs work, there are special power and safety issues involving monitors that don't apply to other parts of the PC. The monitor on a typical PC can consume as much energy as the entire rest of the computer, if not more. The CRT itself produces electromagnetic radiation that can theoretically be harmful to those that spend a great deal of time in front of a PC screen. Because of the tremendous amount of energy consumed by monitors when operating, a couple of initiatives have been started to work on reducing power consumption (and energy use) of monitors during idle periods. The *US's Environmental Protection Agency (EPA)* began a program called *Energy Star* to certify PCs and monitors that meet reduced energy use guidelines. These are sometimes called "Green PCs". Most modern monitors are compliant with *VESA's Display Power Management System* protocol, also called *DPMS. DPMS* is used to selectively shut down parts of the monitor's circuitry after a period of inactivity. With a motherboard and monitor that support *DPMS,* power consumption can be greatly reduced. Motherboards that support *DPMS* often have a *BIOS* setting to enable it. The Operating System or application software used must normally be set to activate *DPMS* after a defined idle period. Many monitors have two low-power settings; a stand-by mode uses less power than the normal operational state, and then an even lower suspend or "shut down" mode turns the monitor off completely to save even more power. The system monitors the PC for activity and after the determined time, sends the appropriate signal to the monitor. When activity is detected again the monitor is "woken up" by the system. One problem with *DPMS* is that if used improperly such as shutting down the system after 1 minute of idle time, can result in a lot of wear and tear on the monitor's internal components, i.e., reducing the monitor's life. If *DPMS* is not used, it is recommended that the monitor should not be left ON for hours at a time if not in use, and especially not unattended overnight. Monitors, unlike other parts of the PC, use very high voltages and have special hazards (mainly electric shock and radiation) that can cause serious injury or even death if a mistake is made while working with one. This is true even if the power is disconnected, due to the large capacitors used to hold charges inside the CRT.

Note: Using a screen saver only, with no other power conservation features, has no significant impact on energy use of the monitor.

o **Phosphor Burned-In:** When electron beams strike the phosphor dots on a CRT, they glow. When a particular image is displayed on a screen for a long time, the electron beam strikes the same dots repeatedly millions of times. If the same exact image is left on some screens for a very long time, it is possible for the surface of the CRT to become damaged. When this happens, "ghosting" can be seen on the surface of the screen, and the outline of the image that was displayed so many times can actually be seen, even when the power to the CRT is OFF. When this happens the phosphor is sometimes said to be "burned-in". The most common place that this phenomenon is seen is at the airport--older monochrome screens that have had the same flight arrival and departure information displayed on them day after day, year after year.

o **Screen Savers:** Screen savers were first invented to address the *Phosphor Burned-In* problem. A screen saver is simply a software program that, after a specified period of inactivity, either blanks the screen or displays a moving pattern on it. This prevents burn in of the screen phosphor that could occur through the same image being on the screen continuously. With the requirement of the *DPMS,* Screen Savers are really unnecessary today. Older monochrome displays were prone to this problem, but it is actually quite rare with modern monitors. Screen savers themselves continue to be popular, but today they are more of a form of entertainment software than a practical utility. Ironically, many screen savers today use their own images that remain stationary on the screen for long periods of time, which means they don't even do what they were originally supposed to do at all. A screen saver is not a replacement for proper power management features. The monitor does not care about what images it is displaying, so it uses power to display the screen saver image as well. If the monitor is using a saver that blanks the screen entirely or is comprised mostly of low-intensity images then slightly less power will be used because the electron gun will be striking the phosphor using less energy, but this still is not the same as using *DPMS,* for example.

o **Electromagnetic Emissions:** All monitors produce emissions as a result of how they work. The electron beam that creates the image also produces electrical and magnetic fields as a side effect. Television sets do the same thing, as they are also based on CRTs. To what extent these emissions are a concern is unknown, and a matter of some controversy in both the computing and health industries. Most agree that the less, the better, but there is no agreement on to what extent emissions can be linked to health problems. In particular, some believe that prolonged exposure to the electromagnetic fields produced by monitors can lead to increased risk of cancer. The Swedish government has been a leading force in developing lower-emissions standards for monitors. The most recent one is called *MPR II,* and many monitors will state specifically if they adhere to this standard. Remember that seeing an advertisement that calls a monitor "lower emissions" is nice but it does not provide all the information to clarify exactly what the word "lower" means. Regardless of the monitor's claims, basic common sense should prevail. It is preferable not to be very close to the CRT for any length of time.

- *Video Adapter card*

A Video Adapter card is a board that plugs into a Personal Computer to give it the display capabilities. The display capabilities of a PC, however, depend on both the logical circuitry (provided in the video adapter) and the display monitor. A Monochrome monitor, for example, cannot display colors no matter how powerful the video adapter. Many different types of video adapters are available for PCs. Most conform to one of the Video Standards defined by *IBM* or *VESA*. Each adapter offers several different video modes. The two basic categories of video modes are *Text* and *Graphics*. In *Text mode*, a monitor can display only ASCII characters. In *Graphics mode*, a monitor can display any bit-mapped image. Within the text and graphics modes, some monitors also offer a choice of resolutions. At lower resolutions a monitor can display more colors. Modern video adapters contain on-board memory, so that the computer's RAM memory is not used for storing displays. In addition, most adapters have their own graphics co-processor for performing graphics calculations. Displays for personal computers have steadily improved since the days of the monochrome monitors that were used in word processors and text-based computer systems in the 1970s. These adapters are often called *Graphics Accelerators*. Video adapters are also called *video cards, video boards, video display boards, graphics cards* and *graphics adapters.*

There are different type of video adapters:

- o *Monochrome Display Adapter (MDA):* The first video cards used in the earliest machines conformed to the *MDA* standard, established by *IBM* as part of the original PC. *MDA* is a monochrome-only, text-only standard, allowing text display at 80x25 characters. Each character is made up of a matrix that is 9 dots wide by 14 dots high, yielding an effective resolution of "720 x 350" at a refresh rate of 50 Hz (of course it is text-only so these dots are not individually addressable). Obviously, *MDA* is obsolete. However, at the time it was a good solution for the limited capabilities of the original PC.

- o *Color Graphics Adapter (CGA):* In 1981, *IBM* introduced the *Color Graphics Adapter*. The *CGA* was the first graphics adapter that was used in the *IBM PC* microcomputer. This display system was capable of rendering four colors, and had a maximum resolution of 320 pixels horizontally by 200 pixels vertically. While *CGA* was all right for simple computer games such as solitaire and checkers, it did not offer sufficient image resolution for extended sessions of word processing, desktop publishing, or sophisticated graphics applications. The *CGA* was the first mainstream video card to support color graphics on the PC. The *CGA* supports several different modes; the highest quality text mode is "80 x 25" characters in 16 colors. Graphics modes range from monochrome at "640 x 200" to 16 colors at "160 x 200" resolution. The card refreshes at 60 Hz. Note that the maximum resolution of *CGA* is actually significantly lower than *MDA* at "640 x 200" resolution. These dots are accessible individually when in a graphics mode but in text each character was formed from a matrix that is "8 x 8", instead of the *MDA's* "9 x 14" matrix, resulting in much poorer text quality. *CGA* was obsolete, having been replaced by *EGA*. *CGA's* highest-resolution mode is 2 colors at a resolution of 640 by 200. *CGA* was superseded by *EGA* systems.

- o *Enhanced Graphics Adapter (EGA):* In 1984, *IBM* introduced the *Enhanced Graphics Adapter* display. The *EGA* was later introduced for the *IBM PC* to provide users with sharper images on the computer's monitor. It allowed up to 16 different

167

colors and improved the resolution to 640 pixels horizontally by 350 pixels vertically. This improved the appearance of the display and made it possible to read text more easily than with *CGA*. Nevertheless, *EGA* did not offer sufficient image resolution for high-level applications such as graphic design and desktop publishing. The *EGA* offered improved resolutions and more colors than *CGA*, although the capabilities of *EGA* are still quite poor compared to modern devices. *EGA* allowed graphical output up to 16 colors (chosen from a palette of 64) at screen resolutions of 640x350, or 80x25 text with 16 colors, all at a refresh rate of 60 Hz. *EGA* was superseded by *VGA* systems.

o **MultiColor Graphics Array (MCGA):** The *MCGA* was introduced with the *IBM PS/2* to produce even sharper images than the **EGA**.

o **Video Graphics Array (VGA):** The *VGA* was similar to the *MCGA* except it was used for *IBM PC* compatible computers. IBM introduced it in 1987. *VGA* remains the lowest common denominator. All PCs made today support *VGA*, and possibly some other more advanced standard, the replacement for *EGA*. *VGA*, supersets of *VGA*, and extensions of *VGA* form today the basis of virtually every video card used in PCs. Introduced in the IBM PS/2 model line, *VGA* was eventually cloned and copied by many other manufacturers. *VGA* was extended and adapted in many different ways. Most video cards today support resolutions and color modes far beyond what *VGA* really is, but they also support the original *VGA* modes, for compatibility. Most call themselves "*VGA compatible*" for this reason. Many people do not realize just how limited true *VGA* really is; *VGA* is obsolete itself by today's standards, and 99% of people using any variant of Windows are using resolution that exceeds the *VGA* standards. True *VGA* supports 16 colors at "640 x 480" resolution, or 256 colors at "320 x 200" resolution (and *not* 256 colors at "640 x 480"). *VGA* colors are chosen from a palette of 262,144 colors (not 16.7 million) because *VGA* uses 6 bits to specify each color, instead of the 8 that is the standard today. *VGA* (and *VGA* compatibility) is significant in one other way as well: they use output signals that are totally different than those used by older standards. Older displays sent digital signals to the monitor, while *VGA* send analog signals. This change was necessary to allow for more color precision. Older monitors that work with *EGA* and earlier cards use "*Transistor-Transistor Logic*" or "*TTL*" signaling and will not work with *VGA*. Some monitors that were produced in the late 80s actually have a toggle switch to allow the selection of either digital or analog inputs. Note that standard *VGA* does not include any hardware acceleration features: the system processor does all the work of creating the displayed image. All acceleration features are extensions beyond standard *VGA*. *VGA* has become one of the de facto standards for PCs. In text mode, *VGA* systems provide a resolution of 720 by 400 pixels. In graphics mode, the resolution is either 640 by 480 (with 16 colors) or 320 by 200 (with 256 colors). The total palette of colors is 262,144. Unlike earlier graphics standards for PCs -- *MDA*, *CGA*, and *EGA* -- *VGA* uses analog signals rather than digital signals. Consequently, a monitor designed for one of the older standards will not be able to use *VGA*.

o **8514/A:** This standard was actually introduced at the same time as standard *VGA*, and provides both higher resolution/color modes and limited hardware acceleration capabilities as well. By modern standards 8514/A is still rather primitive: it supports

1024x768 graphics in 256 colors but only at 43.5 Hz (*interlaced*), or 640x480 at 60 Hz (*non-interlaced*).

- o **Extended Graphics Array (XGA):** In 1990, *IBM* introduced the *XGA* display as a successor to its *8514/A* display. A later version, *XGA-2* offered "800 by 600" pixel resolution in True Color (16 million colors) and 1,024 by 768 resolution in 65,536 colors. The *XGA* cards were used in later *PS/2* models; they can do Bus mastering on the *MCA Bus* and use either 512 KB or 1 MB of VRAM. In the 1 MB configuration, *XGA* supported 1,024x768 graphics in 256 colors, or 640x480 at high color (16 bits per pixel).

- o **Super Video Graphics Array (SVGA):** Most PC displays sold today are described as *SVGA* displays. *SVGA* originally meant "*beyond VGA*" and was not a standard. More recently, the *Video Electronics Standards Association (VESA)* has established a standard programming interface for *SVGA* displays, called the *VESA BIOS* Extension. Typically, an *SVGA* display can support a palette of up to 16,000,000 colors, although the amount of video memory in a particular computer may limit the actual number of displayed colors to something less than that. Image-resolution specifications vary. In general, the larger the diagonal screen measure of an *SVGA* monitor, the more pixels it can display horizontally and vertically. Small *SVGA* monitors (14-inch diagonal) usually display 800 pixels horizontally by 600 pixels vertically. The largest monitors (20 inches or more diagonal measure) can display "1280 x 1024", or even "1600 x 1200", pixels. *VGA* was the last well-defined and universally accepted standard for video. After *IBM's VGA* definition, many companies came into the market and created new cards with more resolution and color depths than standard *VGA* (but almost always, backwards compatible with *VGA*). Most video cards (and monitors for that matter) today advertise themselves as being *Super VGA (SVGA)*. *SVGA* refers collectively to any and all of a host of resolutions, color modes and poorly accepted pseudo-standards that have been created to expand on the capabilities of *VGA*. In the current world of multiple video standards; resolutions, color depths and refresh rates need to be identified on each card and that the monitor it is connected to support the modes the video card produces. *Super VGA* is a set of graphics standards designed to offer greater resolution than *VGA*. There are several varieties of *SVGA*, each providing a different resolution:
  - o 800 by 600 pixels.
  - o 1024 by 768 pixels
  - o 1280 by 1024 pixels
  - o 1600 by 1200 pixels

  All *SVGA* standards support a palette of 16 million colors, but the number of colors that can be displayed simultaneously is limited by the amount of video memory installed in a system. One *SVGA* system might display only 256 simultaneous colors while another displays the entire palette of 16 million colors. A consortium of monitor and graphics manufacturers called *VESA* developed the *SVGA* standards.

- o **Beyond SVGA:** Resolution that is higher than the *SVGA* is sometimes called *Ultra VGA* or *UVGA*. Some people like to refer to *VGA* as "640 x 480", SVGA as "800 x

600", and *UVGA* as "1024 x 768" resolution. The proliferation of video chipsets and standards has created the reliance on software drivers that PC users have come to know well. While Microsoft Windows, for example, has a generic *VGA* driver that will work with almost every video card available, using the higher resolution capabilities of the video card requires a specific driver written to work with the card. The *VESA* standards have changed this somewhat, but not entirely. *IBM* did create several new video standards after *VGA* that expanded on its capabilities. Compared to *VGA*, these have received very limited acceptance in the market, mainly because they were implemented on cards that used *IBM's* proprietary *Micro Channel Architecture (MCA)*. These video standards are:

- o *Extended Graphics Array - 2 (XGA-2):* This graphics mode improves on *XGA* by extending 1,024x768 support to high color, and also supporting higher refresh rates than *XGA* or *8514/A*.

Popular Video Standards for PCs

Video Standard	Resolution	Simultaneous Colors
VGA	640 by 480	16
	320 by 200	256
SVGA	800 by 600	16
	1,024 by 768	256
	1,280 by 1,024	256
	1,600 by 1,200	256
8514/A	1,024 by 768	256
XGA	640 by 480	65,536
	1,024 by 768	256

Table 15.4

- • *Text and Graphical Modes*

With the exception of the very earliest cards used on old PCs in the early to mid 80s, all video cards are able to display information in either *text or graphical modes*. In a *text mode*, video information is stored as characters in a character set; usually on PCs this is the ASCII character set. A typical PC text screen has 25 rows and 80 columns. The video card has built into it a definition of what the dot shape is for each character, which it uses to display the contents of the screen. *Graphical mode* is totally different; the dots on the screen are manipulated directly, so both text and images are possible. The conversion of letters, numbers etc. to visible images is done by software. This is the concept behind fonts; open the same file and display it under a different font and the appearance is totally different. Graphical modes allow for much more flexibility in terms of what is displayed on the screen: they require much more information to be manipulated, and also much more memory to hold the screen image. The increase is significant: typically a factor of up to 100 times or more. This has led almost directly to the need for increased hardware power in new PCs. Most PCs use both text and graphical modes, and can be switched between them under software control. While most computing is now done in a graphics mode, DOS is still text-based. PCs also generally boot up in a text or text-emulated mode.

- *Refresh Rates and Interlacing*

The *RAMDAC* is the device in the video card that is responsible for reading the contents of the video memory, converting the digital values in memory into analog video signals, and sending them over the video cable to the monitor. The *RAMDAC*'s ability to translate and transfer this information directly controls the *refresh rate* for the video mode it is operating in. The refresh rate is the number of times per second that the *RAMDAC* is able to send a signal to the monitor and the monitor is able to repaint the screen. Refresh rate is measured in Hertz (Hz), a unit of frequency. Support for a given refresh rate requires two things: a video card capable of producing the video images that many times per second, and a monitor capable of handling and displaying that many signals per second. The refresh rates are somewhat standardized; common values are 56, 60, 65, 70, 72, 75, 80, 85, 90, 95, 100, 110 and 120 Hz. This is done to increase the chance of compatibility between video cards and monitors. The refresh rate is not the frame rate. The frame rate of a program refers to how many times per second the graphics engine can calculate a new image and put it into the video memory. The refresh rate is how often the contents of video memory are sent to the monitor. Frame rate is much more a function of the type of software being used and how well it works with the acceleration capabilities of the video card. It has nothing at all to do with the monitor. The refresh rate is important because it directly impacts the view ability of the screen image. Refresh rates that are too low cause annoying *flicker* that can be distracting to the viewer and can cause fatigue and eye strain. The refresh rate necessary to avoid this varies with the individual, because it is based on the eye's ability to notice the repainting of the image many times per second. Note that this also depends on the size of the monitor. Setting the monitor too high resolution can be counter-productive. The reason is that the higher the refresh rate, the faster the electron guns have to switch between colors for adjacent pixels. At very high refresh rates there can theoretically be a reduction in the contrast in the displayed image. The refresh rate is related directly to the resolution of the image--higher resolution images generally have lower refresh rates. Higher refresh rates require the *RAMDAC* to generate the video images more times per second. The ability of the *RAMDAC* to do this depends on several variables:

o *The general quality of the video card and its circuitry:* Some cards are just not designed to go above a certain refresh rate regardless of the other factors.

o *Screen resolution:* Higher resolutions mean more pixels that the *RAMDAC* has to pump out to maintain a given refresh rate.

o *Video memory bandwidth:* The *RAMDAC* has to read from the video memory the information it puts out to the monitor. The faster it can read this information the faster it can display it. Higher refresh rates require more video bandwidth, which itself depends on several factors that are discussed here.

Refresh rates are normally specified for *non-interlaced* operation, since that is what modern video systems typically use. Some older monitors can only display some of the higher resolutions when using *interlacing*. Interlacing allows the refresh rate to be double what it normally would be, by displaying alternating lines on each refresh. In essence, half the screen is redrawn at a time. Interlaced operation is normally done at 87 Hz (really 43.5 Hz because of the interlacing) and hence produces flicker that is noticeable by most people. *Table 15.4* shows the relationship between screen resolutions, refresh rate and the amount of data the *RAMDAC* must process. The numbers in the table are in MHz, representing how many millions of pixels per second the *RAMDAC* must output to support a given resolution at a given refresh rate. Many video cards rate their *RAMDAC* in MHz. *Table 15.5* includes a 1.32 conversion factor to take into account retrace times (the time that the electron guns are in non-visible areas of the monitor):

Resolution	43.5 Hz (87 Interlaced)	60 Hz	72 Hz	80 Hz	85 Hz	90 Hz	100 Hz
320x200	3.7	5.1	6.1	6.8	7.2	7.6	8.4
640x480	17.6	24.3	29.2	32.4	34.5	36.5	40.6
800x600	27.6	38	45.6	50.7	53.9	57	63.4
1024x768	45.2	62.3	74.7	83	88.2	93.4	103.8
1280x1024	75.3	103.8	124.6	138.4	147.1	155.7	173
1600x1200	110.2	152.1	182.5	202.8	215.4	228.1	253.4

Table 15.5

- ### Frame Buffer

The *frame buffer* is the video memory used to hold the video image displayed on the screen. The amount of memory (in Megabytes) required holding the image depends primarily on the resolution of the screen image and the color depth used per pixel. To calculate how much video memory is required at a given resolution and bit depth is:

$$\text{Video Memory} = \frac{(X-resolution * Y-resolution * bits - per - pixel)}{(8 * 1,048,576)}$$

This is the minimum required memory, it is recommended to have more memory than this formula computes. Video cards are only available in certain memory configurations; a card with 1.87 MB of memory would not be available but 2 MB is. Some video cards address memory in 8, 16 or 32 bits at a time; they cannot do 24-bit access to memory. Memory is commonly structured to allow access in amounts of bits that are straight powers of two (2^3, 2^4 or 2^5). When using *true color*, 32 bits of memory is required, the extra 8 bits per pixel are generally wasted. Due to the round up that occurs because video cards use memory amounts in whole megabytes, there is no difference between 24-bit and 32-bit memory for "800 x 600" and "1024 x 768" resolutions. Both require 2 MB of video RAM or 4 MB of video RAM, respectively. *Table 15.6* shows, in binary megabytes, the amount of memory required for the frame buffer for each common combination of screen resolution and color depth. In parentheses, the smallest industry standard video memory configuration required to support the combination is shown, based on conventional video memory technology:

Resolution	4 Bits	8 Bits	16 Bits	24 Bits	32 Bits
320x200	0.03 (256 KB)	0.06 (256 KB)	0.12 (256 KB)	0.18 (256 KB)	--
640x480	0.15 (256 KB)	0.29 (512 KB)	0.59 (1 MB)	0.88 (1 MB)	1.17 (2 MB)
800x600	--	0.46 (512 KB)	0.92 (1 MB)	1.37 (2 MB)	1.83 (2 MB)
1024x768	--	0.75 (1 MB)	1.50 (2 MB)	2.25 (4 MB)	3.00 (4 MB)
1280x1024	--	1.25 (2 MB)	2.50 (4 MB)	3.75 (4 MB)	5.00 (6 MB)
1600x1200	--	1.83 (2 MB)	3.66 (4 MB)	5.49 (6 MB)	7.32 (8 MB)

Table 15.6

CHAPTER XVI
PRINTER

Printout, which is produced by a printer, is a processed data that a human can read. The printout can be in many forms, large or small, multi-color, graphics, text, etc. For the computer system, the printer is another type of output. Output is data that has been processed into information by the computer. Output must be in a form that is convenient for the users. Printers, plotters or display monitors, can produce output. Printers and plotters produce hard copies of output; display monitors produce soft copy. The printer is very common with computers. To produce a hard copy, the printer first receives electronic signals from the CPU of the PC.

The *printer* is a device that prints text or illustrations on paper. Printers are also classified by the following characteristics:

- **Quality of type:** The output produced by printers is identified as:
  - **Letter Quality (LQ):** *LQ* refers to print that has the same quality as that produced by a typewriter. The term *letter quality* is really something of a misnomer now, because *Laser* printers produce print that is considerably better than that produced by a typewriter. Many *Dot matrix* printers produce a high-quality print known as *near letter quality*. A lower classification of print quality is called *draft quality*. Computer printers are sometimes divided into two classes: those that produce letter-quality type, such as *Laser*, *Ink-jet*, and *Daisy wheel* printers; and those that do not, including most *Dot matrix* printers.
  - **Near Letter Quality (NLQ):** *NLQ* is a quality of print that is not quite letter quality, but is better than draft quality. Many *Dot matrix* printers produce near letter quality print.
  - **Draft Quality:** Describes print whose quality is less than *near letter quality*. Most 9-pin *Dot matrix* printers produce draft-quality print.
- **Speed:** Measured in *characters per second (cps)* or *pages per minute (ppm)*, the speed of printers varies widely. *Daisy wheel printers* tend to be the slowest, printing about 30 *cps*. *Line printers* are fastest (up to 3,000 *lines per minute* or *lpm*). *Dot-matrix* printers can print up to 500 *cps*, and *laser printers* range from about 4 to 20 text *ppm*.
- **Graphics:** Some printers (*Daisy wheel* and *Line printers*) can print only text. Other printers can print both text and graphics.
- **Fonts:** Some printers, notably *Dot matrix printers*, are limited to one or a few *fonts*. In contrast, laser and ink-jet printers are capable of printing an almost unlimited variety of *fonts*. *Daisy wheel printers* can also print different *fonts*, but the daisy wheel need to be changed, making it difficult to mix *fonts* in the same document. It is a design for a set of characters. A *font* is the combination of *typeface* and other qualities, such as size, pitch, and spacing. For example, Times Roman is a *typeface* that defines the shape of each character. Within Times Roman, however, there are many *fonts* to choose from such as *italic*, **bold** at different sizes. The height of characters in a *font* is measured in *points*, each *point* being approximately $\frac{1}{72}$ inch. The width is measured by *pitch*, which refers to *how many characters can fit in an inch*. Common *pitch* values are 10 and 12. A *font* is

identified as a *fixed pitch* if every character has the same width. If the widths vary depending on the shape of the character, it is called a *proportional font*.

There are many different types of printers.

- *Impact printer:* The signals activate print elements, which are pressed against the printer's paper. *Impact printer* refers to a class of printers that work by banging a head or needle against an ink ribbon to make a mark on the paper. *Impact printers* come in a variety of shapes and sizes. Some print one character at a time, while others print a line at a time. Impact printers include all printers that work by striking an ink ribbon such as *Dot matrix, Daisy wheel and Line printers*.

  o *Dot matrix: Dot matrix* also called *Wire matrix (Figure 16.0)*, this type of printer is normally used for rough drafts or other cases when the quality of the print is not an important factor. *Dot matrix printer* creates characters by striking pins against an ink ribbon to print closely spaced dots in the appropriate shape. Each pin makes a dot, and combinations of dots form characters and illustrations. *Dot matrix printers* are relatively expensive and do not produce high-quality output. However, they can print to multi-page forms (carbon copies), something laser and ink-jet printers cannot do. Dot-matrix printers vary in two important characteristics:

Figure 16.0

  - *Speed:* The *speed* can vary from about 50 to over 500 *characters per second (cps)*. Most *Dot matrix printers* offer different speeds depending on the quality of print desired.
  - *Print quality:* Determined by the number of pins (the mechanisms that print the dots), the *print quality* can vary from 9 to 24. The best *Dot matrix printers* (24 pins) can produce near letter-quality type.

  o Daisy wheel: Similar to a ball-head typewriter, this type of printer has a plastic or metal wheel on which the shape of each character stands out in relief. A hammer presses the wheel against a ribbon, which in turn makes an ink stain in the shape of the character on the paper. Daisy wheel printers produce letter-quality print but cannot print graphics. The daisy wheel is a disk made of plastic or metal on which characters stand out in relief along the outer edge. To print a character, the printer rotates the disk until the desired letter is facing the paper. Then a hammer strikes the disk, forcing the character

175

to hit an ink ribbon, leaving an impression of the character on the paper. The daisy wheel can be changed to print different fonts. *Daisy wheel* printers cannot print graphics, and in general they are noisy and slow, printing from 10 to about 75 characters per second. As the price of laser and ink-jet printers has declined, and the quality of *Dot matrix* printers has improved, daisy-wheel printers have become obsolete.

o **Line printer:** The *Line printer* is a high-speed printer capable of printing an entire line at one time. It contains a chain of characters or pins that print an entire line at one time. Line printers are very fast, but produce low-quality print. A fast line printer can print as many as 3,000 lines per minute. The disadvantages of *Line printers* are that they cannot print graphics, the print quality is low, and they are very noisy.

o **Keyboard printer:** This printer is similar to an office typewriter. All instructions, including spacing, carriage return and printing characters are sent from the PC's CPU to the printer. The keyboard allows the user to communicate with the PC. *Keyboard printer* is relatively a slow printer.

o **Chain printer:** The *Chain printer* has a character set assembled in a chain that revolves horizontally past all print positions. The printer has one print hammer for each column on the paper. Characters are printed when hammers press the paper against an inked ribbon. which in turn presses against appropriate characters on the print chain. The fonts can be changed easily on chain printers, allowing a variety of fonts, such as *italic* or **boldface**, to be used. Some chain printers can print up to 2,000 lines per minute.

o **Drum printer:** This printer uses a metal cylinder with rows of characters engraved across its surface. Each column on the drum contains a complete character set and corresponds to one print position on the line. As the drum rotates, all characters are rotated past the print position. A hammer presses the paper against the ink ribbon and drum when the appropriate character is in place. One line is printed for each revolution of the drum, since all characters eventually reach the print position during one revolution. Some drum printers can print 2,000 lines per minute.

• **Non-impact printer:** This type of printer use heat, laser technology or photographic techniques to print output. *Non-impact printer* is a type of printer that does not operate by striking a head against a ribbon. Non-impact printers include laser printers and ink-jet printers. The term *non-impact* is important primarily in that it distinguishes quiet printers from noisy (impact) printers.

o Laser printer: The Laser printer (Figure 16.1) is type of printer that utilizes a laser beam to produce an image on a drum. The light of the laser alters the electrical charge on the drum wherever it hits. The drum is then rolled through a reservoir of toner, which is picked up by the charged portions of the drum. Finally, the toner is transferred to the paper through a combination of heat and pressure. This is also the way copy machines work. Because an entire page is transmitted to a drum before the toner is applied, *laser printers* are sometimes called *page printers*. One of the chief characteristics of laser printers is their resolution -- how many dots per inch (dpi) they lay down. The available resolutions range from 300 dpi at the low end to 1,200 dpi at the high end.

Some laser printers achieve higher resolutions with special techniques known generally as *resolution enhancement*. In addition to the standard monochrome laser printer, which uses a single toner, there also exist color laser printers that use four toners to print in full color. *Color laser printers* are very expensive. *Laser printers* produce very high-quality print and are capable of printing an almost unlimited variety of fonts. Most laser printers come with a basic set of fonts, called *internal* or *resident fonts*, but additional fonts can be added in one of two ways:

Figure 16.1

- *font cartridges:* Laser printers have slots in which font cartridges can be inserted, ROM boards on which fonts have been recorded. The advantage of font cartridges is that they use none of the printer's memory.
- *soft fonts:* All laser printers come with a certain amount of RAM memory, and usually the amount of memory can be increased by adding memory boards in the printer's expansion slots. Fonts, then, can be copied from a disk to the printer's RAM. This is called *downloading* fonts. A font that has been downloaded is often referred to as a *soft font*, to distinguish it from the *hard fonts* available on font cartridges. The more RAM a printer has, the more fonts that can be downloaded at one time.

In addition to text, laser printers are very adept at printing graphics. However, significant amounts of memory is needed in the printer to print high-resolution graphics. To print a full-page graphic at 300 dpi, for example, at least 1 MB is needed of printer RAM. For a 600-dpi graphic, at least 4 MB RAM is needed. Because laser printers are *nonimpact* printers, they are much quieter than dot matrix or daisy wheel printers. They are also relatively fast, although not as fast as some dot-matrix printers. The speed of laser printers ranges from about 4 to 20 *pages of text per minute (ppm)*. A typical rate of 6 ppm is equivalent to about 40 *characters per second (cps)*. Laser printers are

177

controlled through *page description languages (PDLs)*. There are two standards for *PDLs*:

- **PCL:** Hewlett-Packard (HP) was one of the pioneers of laser printers and has developed a *Printer Control Language (PCL)* to control output. There are several versions of *PCL*, so a printer may be compatible with one but not another. In addition, many printers that claim compatibility cannot accept HP font cartridges.

- **PostScript:** This is the standard for Apple Macintosh printers and for all desktop publishing systems.

  Most software can print using either of these *PDLs*. PostScript tends to be a bit more expensive, but it has some features that *PCL* lacks and it is the standard for desktop publishing. Some printers support both *PCL* and PostScript.

There are two other types of page printers that fall under the category of *laser printers* even though they do not use lasers at all:

- **Light-Emitting Diodes (LED):** An array of *LED* is used to expose the drum. The *LED* is an electronic device that lights up when electricity is passed through it. *LEDs* are usually red. They are good for displaying images because they can be relatively small, and they do not burn out. However, they require more power than *LCDs*.

- **Liquid Crystal Display (LCD):** *LCD* is a type of display used in digital watches and many portable computers. *LCD* displays utilize two sheets of polarizing material with a liquid crystal solution between them. An electric current passed through the liquid causes the crystals to align so that light cannot pass through them. Each crystal, therefore, is like a shutter, either allowing light to pass through or blocking the light. Monochrome *LCD* images usually appear as blue or dark gray images on top of a grayish-white background. *Color LCD* displays use two basic techniques for producing color: *Passive matrix* is the less expensive of the two technologies. The other technology, called *thin film transistor (TFT)* or *Active-matrix*, produces color images that are as sharp as traditional *CRT* displays, but the technology is expensive. Recent passive-matrix displays using new *CSTN* and *DSTN* technologies produce sharp colors rivaling active-matrix displays. Most *LCD* screens used in notebook computers are backlit to make them easier to read. Instead of using a laser to create an image on the drum, however, it shines a light through a liquid crystal panel. Individual pixels in the panel either let the light pass or block the light, thereby creating an image composed of dots on the drum. Liquid crystal shutter printers produce print quality equivalent to that of laser printers. Once the drum is charged, however, they both operate like a real laser printer. By comparison, offset printing usually prints at 1,200 or 2,400 *dpi*.

  o **Ink-Jet printer:** This is a type of printer that works by spraying ionized ink at a sheet of paper. Magnetized plates in the ink's path direct the ink onto the paper in the desired shapes. *Ink-jet printers (Figure 16.2)* are capable of producing high quality print approaching that produced by laser printers. A typical *Ink-jet printer* provides a resolution of 300 dots per inch, although some newer models offer higher resolutions. In general, the price of ink-jet printers is lower than that of laser printers. However, they are also considerably slower. Another drawback of *Ink-jet printers* is that they require a special type of ink that is apt to smudge on inexpensive copier paper. Because *Ink-jet printers* require smaller mechanical parts than laser printers, they are especially popular as portable printers. In addition, *color Ink-jet printers* provide an inexpensive way to print full-color documents.

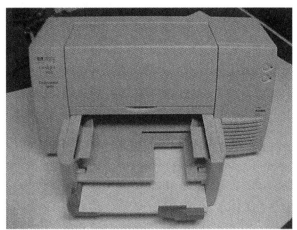

Figure 16.2

o **Thermal printer:** A type of printer that produces images by pushing electrically heated pins against special heat-sensitive paper. Thermal printers are inexpensive and are used in most calculators and some low-end fax machines. They produce low-quality print, and the paper tends to curl and fade. It is an inexpensive printer that works by pushing heated pins against heat-sensitive paper. *Thermal printers* are widely used in calculators and fax machines.

o **Bilevel printer:** A type of printer that can print only two levels of intensity for each dot - ON or OFF. For monochrome printers, lightness and darkness (shading) is simulated through dithering. *Bilevel color printers* use dithering to produce a wide variety of colors. Most *Ink-jet* and *Laser printers* are *bilevel*. In contrast, expensive color printers, such as *thermal dye printers,* can apply ink at various levels of intensity. Such printers are called continuous-tone printers. A third type of printer, called a *contone* or *multilevel printer* can print at a few intensity levels (usually 8), but not as many as a true continuous-tone printer (usually 256). These printers use a combination of dithering and multi-level printing to produce different colors.

o **Bubble-Jet printer:** This is a type of ink-jet printer developed by *Canon.* The principal difference between bubble-jet printers and other ink-jet printers is that *bubble-jet printers* use special heating elements to prepare the ink whereas *ink-jet printers* use piezoelectric crystals.

o **Host-Based printer:** This is a printer that relies on the host computer's processor to generate printable pages. Most *host-based printers* on the market today use the *GDI* interface built into Windows. Because they do not need a powerful processor of their own, host-based printers tend to be less expensive than conventional printers. But because they share the computer's processor, they may be slow and they may slow down the other computer work. How fast these printers operate depends on how powerful the host computer is and how occupied it is with other operations.

o **Graphical Device Interface (GDI) printer:** It is a printer that has built-in support for Windows *Graphical Device Interface*. *GDI* is used by most Windows applications to display images on a monitor, so when printing from a Windows application to a *GDI printer*, there is no need to convert the output to another format such as *PostScript* or *PCL*. *GDI printers* are sometimes called *host-based printers* because they rely on the host computer to rasterize pages.

o **Electrostatic printer:** This printer forms an image of a character on special paper using a dot matrix of charged wires and pins. The paper is moved through a solution containing ink particles that have charge opposite that of the pattern. The ink particles adhere to each charged pattern of the paper, forming a visible image of each character. *Electrostatic printer* operates quietly.

o **Electro thermal printer:** This printer generates characters by using heat and heat-sensitive paper. Rods are heated as matrix. As the ends of the selected rods touch the heat-sensitive paper, an image is created. *Electro thermal printer* operates quietly. Some *Electrostatic* and *Electro thermal printers* are capable of printing 5,000 lines per minute. They are often used in applications where noise may be a problem such as hospital.

o **Xerographic printer:** This printer is used for printing methods much like those used in common xerographic copying machines. For example, Xerox, the pioneer of this type of printing, has one model that prints on a single sheet (such as 8 ½-by-11 inch size) of plain paper rather than on the continuous form paper normally used. Xerographic printer operates at speed of up to 4,000 lines per minute.

The distinction is important because impact printers tend to be considerably noisier than non-impact printers but are useful for multipart forms such as invoices. In summary, *Table 16.0* shows the different printers and their speed (*cpm = character per minute; lpm = lines per minute*):

Printer Performance	
Printer	**Speed**
Keyboard-printer	Up to 400 *cpm*
Dot matrix	Up to 900 *cpm*
Daisy wheel	Up to 3,000 *cpm*
Chain	Up to 2,000 *cpm*
Drum	Up to 2,000 *cpm*
Electrostatic	Up to 5,000 *lpm*
Electro thermal	Up to 5,000 *lpm*
Laser	Up to 21,000 *lpm*
Xerographic	Up to 4,000 *lpm*
Ink-Jet	Up to 12,000 *cpm*

Table 16.0

• **PostScript:** A *Page Description Language (PDL)* developed by Adobe Systems. PostScript is primarily a language for printing documents on laser printers, but it can be adapted to produce images on other types of devices. PostScript is the standard for desktop publishing because image setters, the very high-resolution printers used by service bureaus to produce camera-ready copy, support it. PostScript is an *object-oriented language*, meaning that it treats images, including fonts, as collections of geometrical objects rather than as bit maps. PostScript fonts are called *outline fonts* because the outline of each character is defined. They are also called *scalable fonts* because their size can be changed with PostScript commands. Given a single typeface definition, a PostScript printer can thus produce a multitude of fonts. In contrast, many non-PostScript printers represent fonts with bit maps. To print a bit-mapped typeface with different sizes, these printers require a complete set of bit maps for each size. The principal advantage of object-oriented (*vector*) graphics over bit-mapped graphics is that object-oriented images take advantage of high-resolution output devices whereas bit-mapped images do not. A PostScript drawing looks much better when printed on a 600-dpi printer than on a 300-dpi printer. A bit-mapped image looks the same on both printers. Every PostScript printer contains a built-in interpreter that executes PostScript instructions. If the laser printer does not come with PostScript support, a cartridge that contains PostScript may be purchased. There are three basic versions of PostScript: Level 1, Level 2 and PostScript 3. Level 2 PostScript, which was released in 1992, has better support for color printing. PostScript 3, release in 1997, supports more fonts, better graphics handling, and includes several features to speed up PostScript printing.

- **Object-Oriented Programming (OOP):** *OOP* refers to a special type of programming that combines data structures with functions to create re-usable objects. *Object-oriented graphics* is the same as *vector graphics*. Otherwise, the term *object-oriented* is generally used to describe a system that deals primarily with different types of objects, and where the actions can be taken depend on what type of object is manipulated. For example an object-oriented draw program might enable the user to draw many types of objects, such as circles, rectangles, triangles, etc. Applying the same action to each of these objects, however, would produce different results. If the action is *Make 3D*, for instance, the result would be a sphere, box, and pyramid, respectively.

- **Printer Control Language (PCL):** *PCL* or *Page Description Language (PDL)* developed by Hewlett Packard and used in many of their laser and ink-jet printers. *PCL 5* and later versions support a *scalable font* technology called *Intellifont*.

- **MultiFunction Peripheral (MFP):** *MFP* or *multifunction printers* is a single device that serves several functions, including printing. Typically, *MFP* can act as a printer, a scanner, a fax machine and a photocopier. These devices are becoming a popular option for users because they are less expensive than buying three or four separate devices. The downsides to combining all these functions in one device are:
  - ➢ If the device breaks, all of its functions are lost at the same time.
  - ➢ The device can do one operation at a time. For example, the device cannot print a document and receive a fax simultaneously.

- **Dithering:** Creating the illusion of new colors and shades by varying the pattern of dots. Newspaper photographs, for example, are dithered; varying the patterns of black and white dots produces different shades of gray. There are no gray dots at all. The more dither patterns that a device or program supports, the more shades of gray it can represent. In printing, dithering is usually called *half toning,* and shades of gray are called *halftones.* Note that *dithering* differs from *gray scaling (Figure 16.3).* In gray scaling, each individual dot can have a different shade of gray.

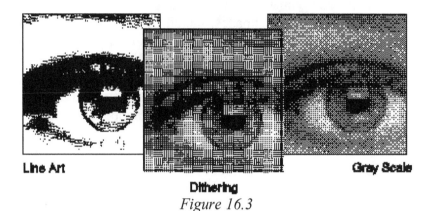

Line Art   Dithering   Gray Scale

Figure 16.3

- **Gray Scaling:** The use of many shades of gray to represent an image. *Continuous-tone* images, such as black-and-white photographs, use an almost unlimited number of shades of gray. Conventional computer hardware and software, however, can only represent a limited number of shades of gray (typically 16 or 256). *Gray scaling* is the process of converting a continuous-tone image to an image that a computer can manipulate. While *gray scaling* is an improvement over monochrome, it requires larger amounts of memory because each dot is represented by from 4 to 8 bits. At a resolution of 300 dpi, more than 8 megabytes is needed to represent a single 8½ by 11-inch page using 256 shades of gray. This can be reduced considerably through data compression techniques, but gray scaling still requires a great deal of memory. Many optical scanners are capable of gray scaling, using from 16 to 256 different shades of gray. However, gray scaling is only useful as an output device – monitor or printer -- that is capable of displaying all the shades. Most color monitors are capable of gray scaling, but the images are generally not as good as on dedicated gray-scaling monitors. Note that gray scaling is different from dithering. Dithering simulates shades of gray by altering the density and pattern of black and white dots. In gray scaling, each individual dot can have a different shade of gray.
- **Plotter:** Sometimes, the best way for a computer to present information to a user is not in form of text but in graphic form, such as a drawing of a house or a manufacturing plant. A *plotter* is an output device that prepares graphic, hard copy of information. It can produce lines, curves and complex shapes. *Plotter* is often used to produce engineering drawings, maps, organizational charts and other large size required printout that may have complex curves, shapes and lines. A typical *plotter* has a pen, a movable cartridge, a drum and a chart-paper holder. Shapes are produced as the pen moves back and forth across the paper along the *y-axis* while the drum moves the paper up and down along the *x-axis*. Both the paper movement and the pen movement are bi-directional. The pen is raised from and lowered to the paper surface automatically. Many *plotters* use more than one pen at a time. Changing the colors of the pens can change the colors of graphics.

CHAPTER XVII

COMMUNICATION

Communication is the ability to pass data and commands from one computer or terminal to another. Communication capability is included in some systems, but requires specialized hardware and software for others. In general, hardware controls the electrical signals connecting systems, while software controls the system so that a document is transferred without errors. In some operating systems, communication facilities are generalized to permit transmission of data across a communication link such as *Private Branch exchange (PBX),* standard telephone line or microwave link. Typically, in a computer system, there are different types of communication: serial, parallel, *USB* and network are the most used.

Communication is the transmission of data from one computer to another, or from one device to another. A communication device is any machine that assists data transmission. For example, modems, cables, and ports are all communications devices. Communications software refers to programs that make it possible to transmit data.

- *Standard*

Standard is a definition or format that has been approved by a recognized standards organization or is accepted as a *"de facto"* standard by the industry. In the PC world, standards exist for programming languages, Operating Systems, data formats, communications protocols, and electrical interfaces. From a user's standpoint, standards are extremely important in the computer industry because they allow the combination of products from different manufacturers to create a customized system. Without standards, only hardware and software from the same company could be used together. In addition, standard user interfaces can make it much easier to learn how to use new applications. Most official computer standards are set by one of the following organizations:

- *American National Standards Institute (ANSI)*

- *Electronic Industries Association (EIA)*

- *International Telecommunication Union (ITU)*

- *Institute of Electrical and Electronic Engineers (IEEE)*

- *International Standards Organization (ISO)*

- *Video Electronics Standards Association (VESA)*

IEEE sets standards for most types of electrical interfaces. Its most famous standard is probably *RS-232C*, which defines an interface for serial communication. This is the interface used by most modems, and a number of other devices, including display screens and mouse. *IEEE* is also responsible for designing floating-point data formats. While *IEEE* is generally concerned with hardware, *ANSI* is primarily concerned with software. *ANSI* has defined standards for a number of programming languages, including *C*, *COBOL*, and *FORTRAN*. *ITU* defines international standards, particularly communications protocols. It has defined a number of standards, including *V.22*, *V.32*, *V.34* and *V.42*, that specify protocols for transmitting data over telephone lines.

183

In addition to standards approved by organizations, there are also *de facto* standards. These are formats that have become standard simply because a large number of companies have agreed to use them. They have not been formally approved as standards, but they are standards nonetheless. *PostScript* is an example of a *de facto* standard.

- **Electronic Industries Association (EIA)**

EIA is a trade association representing the U.S. high technology community. It began in 1924 as the *Radio Manufacturers Association*. The *EIA* sponsors a number of activities on behalf of its members, including conferences and trade shows. In addition, it has been responsible for developing some important standards, such as the *RS-232*, *RS-422* and *RS-423* standards for connecting serial devices.

- **Protocol:** *Protocol* is an agreed-upon format for transmitting data between two devices. The protocol determines the following:

    - The type of error checking to be used

    - Data compression method, if any

    - How the sending device will indicate that it has finished sending a message

    - How the receiving device will indicate that it has received a message

There are a variety of standard protocols from which programmers can choose. Each has particular advantages and disadvantages; for example, some are simpler than others, some are more reliable, and some are faster. From a user's point of view, the only aspect about protocols is that the computer or device must support the right ones to communicate with other computers. The protocol can be implemented either in hardware or in software.

- **Communications Protocols**

Comité Consultatif International Téléphonique et Télégraphique (CCITT): *CCITT* is an organization that sets international communications standards. *CCITT*, now known as *International Telecommunication Union (ITU),* has defined many important standards for data communications (*Table 17.0*), including the following:

Protocol Communications		
Protocol	Maximum Transmission Rate	Duplex Mode
Bell 103	300 bps	Full
CCITT V.21	300 bps	Full
Bell 212A	1,200 bps	Full
ITU V.22	1,200 bps	Half
ITU V.22bis	2,400 bps	Full
ITU V.29	9,600 bps	Half
ITU V.32	9,600 bps	Full
ITU V.32bis	14,400 bps	Full
ITU V.34	36,600 bps	Full
ITU V.90	56,000 bps	Full

Table 17.0

o **Group 3:** The universal protocol, defined by the *CCITT*, for sending fax documents across telephone lines. *Group 3 protocol* specifies *CCITT T.4* data compression and a maximum transmission rate of 9,600 *bits per seconds* or *bps* or *Baud*. There are two levels of resolution: 203 by 98 and 203 by 196.

o **Group 4:** A protocol for sending fax documents over *ISDN* networks. *Group 400* protocol supports images of up to 400 dpi resolution.

o **V.21:** The standard for full-duplex communication at 300 *Baud* in Japan and Europe. In the United States, *Bell 103* is used in place of V.21.

o **V.22:** The standard for half-duplex communication at 1,200 *Baud* in Japan and Europe. In the United States, the protocol defined by *Bell 212A* is more common.

o **V.22bis:** The worldwide standard for full-duplex modems sending and receiving data across telephone lines at 1,200 or 2,400 *Baud*.

o **V.29:** The standard for half-duplex modems sending and receiving data across telephone lines at 1,200, 2,400, 4,800, or 9,600 *Baud*. This is the protocol used by fax modems.

o **V.32:** The standard for full-duplex modems sending and receiving data across phone lines at 4,800 or 9,600 *Baud*. *V.32* modems automatically adjust their transmission speeds based on the quality of the lines.

o **V.32bis:** The *V.32* protocol extended to speeds of 7,200, 12,000, and 14,400 *Baud*.

o **V.34:** The standard for full-duplex modems sending and receiving data across phone lines at up to 28,800 *Baud*. *V.34* modems automatically adjust their transmission speeds based on the quality of the lines.

o **V.42:** An error-detection standard for high-speed modems. *V.42* can be used with digital telephone networks.

o **V.42bis:** A data compression protocol that can enable modems to achieve a data transfer rate of 34,000 *Baud*.

o **V.90**: The standard for full-duplex modems sending and receiving data across phone lines at up to 56,600 *Baud*. In 1998, the *V.90* was approved by *ITU* as a standard for 56 *KBaud* modems. The *V.90* standard resolves the battle between the two competing 56 *KBaud* technologies: *X2* from *3COM* and *K56flex* from Rockwell Semiconductor. Both manufacturers have announced that their future modems will conform to *V.90*.

o **X2:** A technology developed by U.S. Robotics (now *3COM*) for delivering downstream transmission of data rates up to 56K *Baud* over *Plain Old Telephone Service (POTS)*. It was long believed that the maximum data transmission rate over copper telephone wires was 33.6 *KBaud*, but *X2* achieves higher rates by taking advantage of the fact that most phone switching stations are connected by high-speed digital lines. *X2* bypasses the normal digital-to-analog conversion and sends the digital data over the telephone wires directly to the modem where it is decoded. *3COM* has announced that future *X2* modems will conform to the new *V.90* standard approved by the *ITU*. The 56 *KBaud* speed is achieved in the downstream direction only to the PC. Upstream speed is at the regular maximum speed of 33.6 *KBaud*. *X2* provided input to and has been replaced by the *V.90* *ITU* standard. 56K *Baud* technologies exploit the fact that most telephone company offices are interconnected with digital lines. A *V.90*-equipped modem does not need to *demodulate* the downstream data. Instead, it decodes a stream of multi-bit voltage pulses generated as though the line was equipped for digital information. Upstream data still requires digital-to-analog modulation. Unlike *ISDN*, the *V.90* technology does not require any additional installation or extra charges from the phone company. On the other hand, the maximum transmission speed of *Integrated Services Digital Network (ISDN)* is twice that of *V.90* at 128 *Kbaud* which has the flexibility of combining digital and voice transmission on the same line. While *X2* offers faster Internet access than normal modems, there are several caveats to using an *X2* modem:

1. The high speeds are available only with downstream traffic (e.g., data sent to the computer). Upstream traffic is delivered using normal techniques, with a maximum speed of 33.6 *KBaud*.

2. To connect to the Internet at *X2* speeds, the *Internet Service Provider (ISP)* must have a modem at the other end that supports *V.90*.

3. Even if the *ISP* supports *V.90*, the maximum transmission rates might not be achieved due to noisy lines.

o **K56flex:** A technology developed by *Lucent Technologies* and *Rockwell International* for delivering data rates up to 56 *KBaud* over *POTS*. *Lucent* and *Rockwell* have announced that future *K56flex* modems will conform to the new *V.90* standard approved by the *ITU*. And users with older *K56flex* modems may upgrade their modems to support *V.90*. While *K56flex* offers faster Internet access than normal modems, the *X2* caveats are the same when using a *K56flex* modem.

o **X.25:** The most popular packet-switching protocol for LANs. Ethernet, for example, is based on the *X.25* standard.

o **X.400:** The universal protocol for e-mail. *X.400* defines the envelope for e-mail messages so all messages conform to a standard format.

o **X.500:** An extension to *X.400* that defines addressing formats so all e-mail systems can be linked together.

• ***Serial Communication Protocol***

All communications between devices require that the devices agree on the format of the data. The set of rules defining a format is called a *Protocol*. At the very least, a communications protocol must define the following:

- Rate of transmission (in *Baud* or *bits per second* or *bps*).

- *Synchronous* or *Asynchronous* transmission.

- Data is to be transmitted in *Half-duplex* or *Full-duplex* mode.

In addition, *Protocol* can include sophisticated techniques for detecting and recovering from transmission errors and for encoding and decoding data. *Table 17.0* lists the most commonly used protocols for communications via modems. These protocols are almost always implemented in the hardware; they are built into modems. There are a number of protocols that complement these standards (listed in *Table 17.0*) by adding additional functions such as file transfer capability, error detection and recovery, and data compression. The known protocols are *Xmodem, Kermit, MNP*, and *CCITT V.42*. These protocols can be implemented either in hardware or software.

• ***Data Terminal Equipment (DTE)***

DTE is a device that controls data flowing to or from a computer. The term is most often used in reference to serial communications defined by the *RS-232C* standard. This standard defines the two ends of the communications channel as being a *DTE* and *Data Communications Equipment (DCE)* device. In practical terms, the *DCE* is usually a modem and the *DTE* is the computer itself. With internal modems, the *DCE* and *DTE* are part of the same device.

• ***Data Communications Equipment (DCE)***

DCE is a device that communicates with a *Data Terminal Equipment (DTE)* device in *RS-232C* communications.

• ***Universal Asynchronous Receiver-Transmitter (UART)***

The *UART* is a computer component that handles asynchronous serial communication. Every computer contains a *UART* to manage the serial ports, and all internal modems have their own *UART*. As modems have become increasingly fast, the *UART* has come under greater scrutiny as the cause of transmission bottlenecks. The new *UART* contains buffers enabling it to support higher transmission rates.

• ***Asynchronous***

Not synchronized; that is, not occurring at predetermined or regular intervals. The term *asynchronous* is usually used to describe communications in which data can be transmitted intermittently rather than in a steady stream. For example, a telephone conversation is asynchronous because both parties can talk whenever they like. If the communication were synchronous, each party would be required to wait a specified interval before speaking. The difficulty with asynchronous communications is that the receiver must have a way to distinguish between valid data and noise. In computer communications, this is usually accomplished through a special *start bit* and *stop bit* at the beginning and end of each piece of data. For this reason, asynchronous communication is sometimes called *start-stop transmission*. Most communications between computers and devices are asynchronous. *Figure 17.0* is an illustration of a typical serial asynchronous transmission.

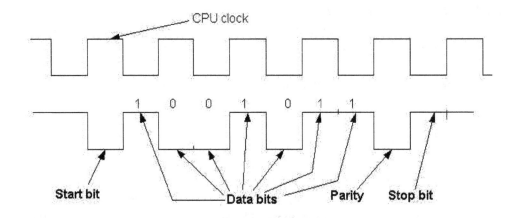

Figure 17.0

- ### *Synchronous*
Synchronous is occurring at regular intervals. The opposite is asynchronous. Most communication between computers and devices is asynchronous because it can occur at any time and at irregular intervals. Communication within a computer, however, is usually synchronous and is governed by the microprocessor's clock. Signals along the bus, for example, can occur only at specific points in the clock cycle.

- ### *Full-duplex (FDX)*
FDX refers to the transmission of data in two directions simultaneously. For example, a telephone is a *full-duplex* device because both parties can talk at once. Most modems have a switch to choose between *full-duplex* and *half-duplex* modes. In *full-duplex* mode, data transmitted does not appear on the monitor until it has been received and sent back by the other party. This validates that the data has been accurately transmitted. If the display screen shows two of each character, it probably means that the modem is set to *half-duplex* mode when it should be in *full-duplex* mode. *Figure 17.1* shows a *Full-duplex* configuration of *DTE* and *DCE* devices. In *Full-duplex* mode, transmitted data is not displayed on the monitor until it has been received and returned (remotely echoed) by the other device.

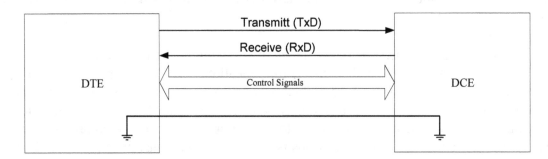

- ### *Half-duplex (HDX)*

Figure 17.1

HDX refers to the transmission of data in just one direction at a time. For example, a *walkie-talkie* is a *Half-duplex* device because only one party can transmit at a time. In *Half-duplex* mode, each character transmitted is immediately displayed on the display screen, it is *local echo*, characters are echoed by the local device.

- *Communication Software*

Software that makes it possible to send and receive data over telephone lines through modems.

- o *Kermit:* This is a communication *protocol* and a set of associated software utilities developed at Columbia University. *Kermit* can be used to transfer files or for terminal emulation. It is frequently used with modems and supports communications via other transport mechanisms such as *TCP/IP*. *Kermit* is noted for its transmission accuracy and slow transmission speeds due to its default settings that optimize for accuracy. However, *Kermit* can also be tuned to transfer data as quickly as any other data transfer protocol. *Kermit* is not in the public domain, but Columbia University allows people to use the protocol for free and almost all communications products support it. However, not all implementations support the full protocol. This has led some people to refer to an advanced version of *Kermit* as *Super Kermit*. Actually, there is only one version of the *Kermit* protocol, which supports all the advanced features usually attributed to *Super Kermit*, such as sliding windows and long packets. Other file transfer protocols used by modems include *Xmodem* and *Zmodem*.

- o *Xmodem:* Developed in 1977 by *Ward Christiansen, Xmodem* is one of the most popular file transfer protocols. *Xmodem* is a relatively simple protocol, it is fairly effective at detecting errors. It works by sending blocks of data together with a checksum and then waiting for acknowledgment of the block's receipt. The waiting slows down the rate of data transmission considerably, but it ensures accurate transmission. *Xmodem* can be implemented either in software or in hardware. Many modems, and almost all communications software packages, support *Xmodem*. However, it is useful only at relatively slow data transmission speeds (4,800 *Baud* or less). With *Xmodem*, only one file can be sent at a time. Enhanced versions of *Xmodem* that work at higher transmission speeds are known as *Ymodem* and *Zmodem*.

- o *Ymodem:* *Ymodem* is an asynchronous communications protocol designed by *Chuck Forsberg* that extends *Xmodem* by increasing the transfer block size and by supporting batch file transfers. With *Ymodem*, a list of files can be sent at one time.

- o *Zmodem:* *Zmodem* is an asynchronous communications protocol that provides faster data transfer rates and better error detection than *Xmodem*. In particular, *Zmodem* supports larger block sizes and enables the transfer to resume where it left off following a communications failure.

- o *Checksum:* A simple error-detection scheme in which each transmitted message is accompanied by a numerical value based on the number of set bits in the message. The receiving station then applies the same formula to the message and checks to make sure the accompanying numerical value is the same. If not, the receiver can assume that the message has been garbled. *Zmodem*, uses *checksum*.

- o *Error-Correcting Code memory (ECC):* *ECC* is a type of memory that includes special circuitry for testing the accuracy of data as it passes in and out of memory.

- o *Cyclic Redundancy Check (CRC):* *CRC* is a common technique for detecting data transmission errors. *CRC* is an error detection scheme in which the block check character is the remainder after dividing all the serialized bits in a transmission block by a

189

predetermined binary number. A number of file transfer protocols, including *Zmodem*, use CRC in addition to *checksum*.

o   **Data Compression:** *Data Compression* is storing data in a format that requires less space than usual. *Compressing* data is the same as *packing* data. Data compression is particularly useful in communications because it enables devices to transmit the same amount of data in fewer bits. There are a variety of data compression techniques, but only a few have been standardized. The *CCITT* has defined a standard data compression technique for transmitting faxes (*Group 3* standard) and a compression standard for data communications through modems (*CCITT V.42bis*). In addition, there are file compression formats, such as *ARC* and *ZIP*. Data compression is also widely used in backup utilities, spreadsheet applications, and database management systems. Certain types of data, such as bit-mapped graphics, can be compressed to a small fraction of their normal size.

o   **Parity:** In computers, parity (from the Latin *paritas*: equal or equivalent) refers to a technique of checking whether data has been lost or written over when it is moved from one place in storage to another or when transmitted between computers. It is an additional bit, the *parity bit*, added to a group of bits then moved together. This bit is used only for the purpose of identifying whether the bits being moved arrived successfully. Before the bits are sent, they are counted and if the total number of data bits is "*even*", the *parity bit* is <u>set to one</u> so that the total number of bits transmitted will form an *odd* number. If the total number of data bits is already an "*odd*" number, the *parity bit* remains or is set to 0. At the receiving end, each group of incoming bits is checked to see if the group totals to an *odd* number. If the total is *even*, a transmission error has occurred and either the transmission is retried or the system halts and an error message is sent to the user via an error message on the monitor. The description above describes how parity checking works within a computer. Specifically, the *PCI* bus and the I/O Bus controller use the *odd* parity method of error checking. *Parity bit* checking is not an infallible error-checking method since it is possible that two bits could be in error in a transmission, offsetting each other. For transmissions within a personal computer, this possibility is considered extremely remote. In some large computer systems where data integrity is seen as extremely important, three bits are allocated for parity checking. Parity checking is also used in communication between modems. Parity checking can be selected to be *even* (a successful transmission will form an *even* number) or *odd*. Users may also select *no parity*, meaning that the modems will not transmit or check a *parity bit*. When *no parity* is selected, it is assumed that there are other forms of checking that will detect any errors in transmission. *No parity* also usually means that the *parity bit* can be used for data, speeding up transmission. In a modem-to-modem communication, the type of *parity* is co-ordinated by the sending and the receiving modems before the transmission takes place.

•   **Input/Output Address:** *I/O address* is a resource used by virtually every device in the computer. Conceptually, it represents the location in memory that is designated for use by various devices to exchange information between themselves and the PC *(Table 17.0)*.

Address	Description
000-00Fh	DMA controller, channels 0 to 3
040-04Fh	System timers and/or System use
060-06Fh	Keyboard, PS/2 mouse and Speaker
070-07Fh	RTC/CMOS, NMI and/or System use
0C0-0DFh	DMA controller, channels 4-7
0F0-0FFh	Floating point unit (FPU/NPU/Math coprocessor)
130-15Fh	SCSI host adapter
160-16Fh	Quaternary IDE controller, master drive
170-17Fh	Secondary IDE controller, master drive
1E0-1EFh	Tertiary IDE controller, master drive
1F0-1FFh	Primary IDE controller, master drive
200-20Fh	Joystick port, System use
220-22Fh	Sound card
230-23Fh	Network card
250-25Fh	System use, Plug & Play system devices, LPT2
270-27Fh	Sound card, Network card, LPT3, COM4
280-28Fh	COM2 (second serial port)
2E0-2Efh	Sound card (MIDI port)
2F0-2FFh	Tape accelerator card, Secondary IDE controller (slave drive), LPT1
300-30Fh	LPT2 (second parallel port) (monochrome systems)
370-37Fh	Network card, VGA/EGA Video, sound card (FM synthesizer)
380-38Fh	VGA/CGA Video
3C0-3CFh	Tape accelerator card, COM3 (third serial port)
3D0-3DFh	Tertiary IDE controller (slave drive)
3E0-3EFh	Floppy disk controller and/or COM1 (first serial port)
	Tape accelerator card, Primary IDE controller (slave drive)
3F0-3FFh	Floppy disk controller, COM1, tape accelerator card, Primary IDE controller (slave drive)

Table 17.0

- *Memory-Mapped I/O:* In an example of a communications (COM) port that has a modem connected to it. When information is received by the modem, it needs to get this information into the PC. The information needs to be put the data as it is pulled off the phone line. The PC gives each device its own small area of memory to work with. This is called *memory-mapped I/O*. When the modem gets a byte of data it sends it over the COM port, and it shows up in the COM port's designated I/O address space. When the CPU is ready to process the data; to find it, the CPU knows where to look. Later, when This CPU wants to *send* information over the modem, it uses this address again. This is a very simple way of dealing with the problem of information exchange between devices.

191

- ***I/O Address Space Width:*** Unlike *IRQs* and *DMA* channels, which are of uniform size and normally assigned one per device--sound cards use more than one because they are really many devices wrapped into one package--I/O addresses vary in size. The reason is simple: some devices (e.g., network cards) have much more information to move around than others (e.g., keyboards). The size of the I/O address is, in some cases, dictated by the design of the card and compatibility reasons with older devices. Most devices use an I/O address space of 4, 8 or 16 bytes; some use as few as 1 byte and others as many as 32 or more. The wide variance in the size of the I/O addresses can make it difficult to determine and resolve resource conflicts, because often I/O addresses are referred to only by the *first byte* of the I/O address. For example, people may set the network card at 360h, which may seem not to conflict with LPT1 parallel port at address 378h. In fact many network cards take up 32 bytes for I/O; this means they use up 360-37Fh, which totally overlaps with the parallel port (378-37Fh).

- ***Logical Devices:*** Some devices have both a physical address and also a logical name. The two most commonly encountered device types that work this way are serial ports (COM1, COM2, COM3 & COM4) and parallel ports (LPT1, LPT2 & LPT3). Actually, hard disks are labeled this way too, A:, C: etc. The purpose of this logical labeling is to make it easier to refer to devices without having to know their specific addresses. It is much simpler for software to be able to refer to a COM port by name than by an address.

- ***Logical Name Assignment:*** Logical device names are assigned by the system *BIOS* during the *Power-On Self Test (POST),* when the system is booted up. The *BIOS* searches for devices by *I/O address* in a predefined order, and assigns them a logical name *dynamically,* in numerical order. *Table 17.1* shows the normal default assignments for COM ports, in order:

Port	I/O Address	Default IRQ
COM1	3F8-3FFh	4
COM2	2F8-2FFh	3
COM3	3E8-3EFh	4
COM4	2E8-2EFh	3

Table 17.1

For parallel ports. it is slightly more complicated. Originally *IBM* defined different defaults for monochrome-based PCs and for color PCs. All new systems have been color for many years, but even some new systems still put LPT1 at 3BCh. Most new systems have LPT1 at 378-37Fh. *Table 17.2* shows parallel port normal default assignment:

Port	Monochrome Systems	Color Systems
LPT1	3BC-3BFh	378-37Fh
LPT2	378-37Fh	278-27Fh
LPT3	278-27Fh	--

Table 17.2

- ***Problems With Logical Device Names:*** Most of the problems that arise with the use of logical device names occur when devices are added or removed from the system. The most common problem is software that will refuse to work because the logical device name assigned to a physical device has changed, as a result of a device being added to or removed from the system. Most software refers to a device by its name such as "*LPT1*". However, the names are assigned dynamically by the *BIOS* at boot time, when it searches the PC system to see what hardware it has. If the PC originally had "*LPT1*" at 378-37Fh and a new parallel port is added at the address of 3BC-3BFh, then the *new* one will now be *LPT1* and the old port will become *LPT2*. This is because the ports are labeled dynamically based on a predefined search order, and 3BC is looked at first.

- ***Parallel communication***

In the context of the Internet and computing, parallel means more than one event happening at a time. It is usually contrasted with *serial*, meaning only one event happening at a time. In data transmission, the techniques of time division and space division are used, where time separates the transmission of individual bits of information sent serially and space (in multiple lines or paths) can be used to have multiple bits sent in parallel. In the context of computer hardware and data transmission, serial connection, operation, and media usually indicate a simpler, slower operation. Parallel connection and operation indicates faster operation. This indication is not always true since a serial medium (for example, fiber optic cable) can be much faster than a slower medium that carries multiple signals in parallel. A conventional phone connection is generally thought of as a serial line since its usual transmission protocol is serial. Conventional computers and their programs operate in a serial manner, with the computer reading a program and performing its instructions one after the other. However, some computers have multiple processors that divide up the instructions and perform them in parallel.

- ***Enhanced Parallel Port/Enhanced Capability Port (EPP/ECP)***

EPP/ECP is a standard signaling method for bi-directional parallel communication (*Table 17.3*) between a computer and peripheral devices that offers the potential for much higher rates of data transfer than the original parallel signaling methods. *EPP* is for non-printer peripherals. *ECP* is for printers and scanners. *EPP/ECP* are part of *IEEE Standard 1284*, which also specifies support for current signaling methods (including *Centronics*, the *de facto* standard for printer communication) so that both old and new peripherals can be accommodated.

193

Printer Connection			
Parallel Port Pin Definitions			
Pin Number	**Function**	**Pin Number**	**Function**
1	Strobe-	2	Auto Feed-
3	Data bit 0	4	Error-
5	Data bit 1	6	Init-
7	Data bit 2	8	SLCT IN-
9	Data bit 3	10	GND
11	Data bit 4	12	GND
13	Data bit 5	14	GND
15	Data bit 6	16	GND
17	Data bit 7	18	GND
19	ACJ-	20	GND
21	Busy	22	GND
23	PE	24	GND
25	SLCT	26	NC

Table 17.3

The new standard specifies five modes of data transfer. Three of them support the older mono-directional modes (a forward direction method from PC to *Centronics* printer and two reverse direction methods from peripheral to the PC). The fourth and fifth modes, *EPP* and *ECP*, are bi-directional (*half-duplex*) signaling methods, meaning that they are designed for back-and-forth communication. Partly because these are being implemented in hardware, *EPP* and *ECP* will provide much faster data transfer. The first three methods offer an effective data transfer rate of 50 to 100 Kilobytes per second. *EPP* and *ECP* offer rates estimated over 1 Megabytes per second. In order to get the maximum advantage of *EPP/ECP*, both Operating System (or an I/O port controller, or both) and peripheral device must support the standard. Even printers that support *ECP* are limited by the mechanical aspects of printing. Nevertheless, even users of the compatibility modes of *Standard 1284* are also expected to see some benefit in data transfer to and from peripherals. *Microsoft's Windows 95* has built-in support for *IEEE 1284* in its parallel plug-and-play feature.

- ***Line Print Terminal (LPT, LPT1, LPT2, and LPT3)***

LPT (Line Print Terminal) is the usual designation for a parallel port connection to a printer or other device on a personal computer. Most PCs come with one or two *LPT* connections designated as *LPT1* and *LPT2*. Some systems support a third, *LPT3*. Whatever the number, *LPT1* is the usual default. A parallel port for a second printer or other device can be added by adding a parallel port adapter card to the PC. An *LPT* port can be used for an input device such as a scanner or a video camera. Parallel computer connections traditionally used the *Centronics* interface for printer communication.

- *Serial Port*

A port, or *interface*, that can be used for serial communication, in which only 1 bit is transmitted at a time. Most serial ports on personal computers conform to the *RS-232C* or *RS-422* standards. A serial port is a general-purpose interface that can be used for almost any type of device, including modems, mouse, and printers (although most printers are connected to a parallel port).

- *Recommended Standard-232C (RS-232C): RS-232C* is a standard interface approved by the *Electronic Industries Association (EIA)* for connecting serial devices. In 1987, the *EIA* released a new version of the standard and changed the name to *EIA-232-D*. And in 1991, the *EIA* teamed up with *Telecommunications Industry Association (TIA)* and issued a new version of the standard called *EIA/TIA-232-E*. Many people, however, still refer to the standard as *RS-232C,* or just *RS-232*. Almost all modems conform to the *EIA-232* standard and most personal computers have an *EIA-232 port* for connecting a modem or other device.

  In addition to modems, many display screens, mouse, and serial printers are designed to connect to *EIA-232 port*. In *EIA-232* parlance, the device that connects to the interface is *DCE* and the device to which it connects (e.g., the computer) is *DTE*. The *EIA-232* standard supports two types of connectors (*Figure 17.2*): a 9-pin D-type connector (DB-9, *Figure 17.3-a*) and a 25-pin D-type connector (DB-25, *Figure 17.3-b*). The type of serial communications used by PCs requires only 9 pins so either type of connector will work equally well.

  The serial interface port is an input/output channel for serial data communication. This is a connection point on a computer used to connect a serial interface device to the computer system. Serial port is typically identified as a communication (COM) port.

Figure 17.2

Serial communication transmits one bit at a time bi-directional, i.e., there is a Transmit and a Receive lines. Data is transmitted: one bit in one direction (transmit) and one bit in the other direction (receive). Because the data is transferred sometimes at high speed, there is a need for staging memory: *buffer memory. Buffer memory* is a part of the computer's memory that is used to store transmitted and/or received data. Buffer memories are used because a peripheral has higher speed requirements than general system memory can support or because additional processing must be performed on the data. Most typical computers have two serial ports (*RS-232*) that are DB-9 and DB-25 male connectors. Example of serial device is modem, mouse, etc.

RS-232 Interface

RS-232 (EIA Std.) applicable to the 25 pin interconnection of Data Terminal Equipment (DTE) and Data Communications Equipment (DCE) using serial binary data

Pin	Description	EIA CKT	From DCE	To DCE
1	Frame Ground	AA		
2	Transmitted Data	BA		D (Data)
3	Received Data	BB	D	
4	Request to Send	CA		C (Control)
5	Clear to Send	CB	C	
6	Data Set Ready	CC	C	
7	Signal Grnd/Common Return	AB		
8	Rcvd. Line Signal Detector	CF	C	
11	Undefined			
12	Secondary Rcvd. Line Sig. Detector	SCF	C	
13	Secondary Clear to Send	SCB	C	
14	Secondary Transmitted Data	SBA		D
15	Transmitter Sig. Element Timing	DB	T (Timing)	
16	Secondary Received Data	SBB	D	
17	Receiver Sig. Element Timing	DD	T	
18	Undefined			
19	Secondary Request to Send	SCA		C
20	Data Terminal Ready	CD		C
21	Sig. Quality Detector	CG		C
22	Ring Indicator	CE	C	
23	Data Sig. Rate Selector (DCE)	CI	C	
23	Data Sig. Rate Selector (DTE)	CH		C
24	Transmitter Sig. Element Timing	DA		T
25	Undefined			

PC Com Port -232

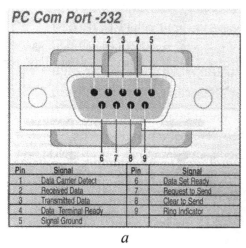

Pin	Signal	Pin	Signal
1	Data Carrier Detect	6	Data Set Ready
2	Received Data	7	Request to Send
3	Transmitted Data	8	Clear to Send
4	Data Terminal Ready	9	Ring Indicator
5	Signal Ground		

a *b*

Figure 17.3

Although *EIA-232* is still the most common standard for serial communication, the *EIA* has defined successors to *EIA-232* called *RS-422* and *RS-423*. The new standards are backward compatible so that *RS-232* devices can connect to an *RS-422* port.

- o **RS-422 and RS-423:** *RS-422 and RS-423* are standard interfaces approved by *EIA* for connecting serial devices. The *RS-422* and *RS-423* standards are designed to replace the older *RS-232* standard because they support higher data rates and greater immunity to electrical interference. Apple Macintosh computers contain an *RS-422* port that can also be used for *RS-232C* communication. *RS-422* supports multipoint connections whereas *RS-423* supports only point-to-point connections.
- o **RS-485:** *RS-485* is *EIA* standard for multipoint communications. It supports several types of connectors, including DB-9 and DB-37. *RS-485* is similar to *RS-422* but can support more nodes per line because it uses lower-impedance drivers and receivers. The *RS-485* standard permits a balanced transmission line to be shared in a party line or multidrop mode. As many as 32 driver/receiver pairs can share a multidrop network. The principle difference between *RS-422* and *RS-485* is that *RS-485* driver can be put into high

impedance, tristate mode which allows other drivers to transmit over the same pair of wires.

- ### *Universal Serial Bus (USB)*

USB is a new communication standard that supports data transfer rates of 12 Mbps (12 Megabits per second). A single *USB* port can be used to connect up to 127 peripheral devices, such as mouse, modems, and keyboards, scanners, storage drives, printers, etc. *USB* also supports *Plug-and-Play* installation (device can be auto-detected and auto-configured by the PC) and *hot plugging (*device can be connected or disconnected without shutting down or restarting the PC). Starting in 1996, few computer manufacturers included *USB* support in their new PC machines. Since the release of Intel's 440LX chipset in 1997, *USB* has become more widespread. If the PC does not have built-in (mainboard), a *PCI* card or an external hub with external power can be added. The new *USB* hub must comply with *USB 1.1* or later. Earlier versions are not reliable and can lose data especially when using data storage. The *USB* access is not fast; it was not designed for speed. It does not work with Microsoft NT 4.0, works with Microsoft Windows 95, Revision B, Windows 98 and Macintosh machines.

- ### *IEEE 1394*

IEEE 1394 is a new, very fast external bus standard that supports data transfer rates of up to 400 Mbps (400 million bits per second). Products supporting the *1394* standard label it different names, depending on the company. Apple, which originally developed the technology, uses the trademarked name *FireWire*. Other companies use other names, such as *I-link* and *Lynx,* to describe their *1394* products. A single *1394* port can be used to connect up 63 external devices. In addition to its high speed, *1394* also supports *isochronous data* -- delivering data at a guaranteed rate. This makes it ideal for devices that need to transfer high levels of data in real-time, such as video devices. Although extremely fast and flexible, *1394* is also expensive. Like *USB*, *1394* supports both *Plug-and-Play* and *hot plugging*, and also provides power to peripheral devices. The main difference between *1394* and *USB* is that *1394* supports faster data transfer rates and is more expensive. For this reason, *1394* is expected mostly for devices that require large throughputs, such as video cameras, whereas *USB* will be used to connect most other peripheral devices.

- ### *Bulletin Board System (BBS)*

BBS is an electronic message center. Most bulletin boards serve specific interest groups. They allow PC users to dial in with a modem; review messages left by others, and leave messages if they opt to. *BBS* are a particularly good place to find free or inexpensive software products. In most cases, *BBS* is being replaced by Internet websites.

- ### *File Transfer Protocol (FTP)*

FTP, a standard protocol, is the simplest way to exchange files between computers on the Internet. Like the *Hypertext Transfer Protocol (HTTP)*, which transfers displayable Web pages and related files, and the *Simple Mail Transfer Protocol (SMTP),* which transfers e-mail, *FTP* is an application protocol that uses the Internet's *TCP/IP* protocols. *FTP* is commonly used to transfer Web page files from their creator to the computer that acts as their server for everyone on the Internet. It is also commonly used to download programs and other files to a computer from other servers. As a user, *FTP* can be used with a simple command line interface or with a software program that offers a graphical user interface. The Web browser can also make *FTP* requests to download selected programs from a Web page. Using *FTP*, files at a server can also be updated (deleted, renamed, moved, or copied).

- ***Fractional T1 or T3 line***

A fractional *T1* (also known as *DS1 - Digital Signal*) or *T3* line is a digital phone line in the North American T-carrier system that is leased to a customer at a fraction of its data-carrying capacity and at a correspondingly lower cost. A *T-1* line contains 24 channels, each with a data transfer capacity of 64K *bps*. The customer can rent any number of the 24 channels. The transmission method and speed of transfer remain the same (*T1* is about 1.544 *Megabits per second*). Overhead bits and framing are still used, but the un-rented channels simply contain no data. *T3* lines (which offer 672 channels at 64K *bps* each) are also sometimes offered as a fractional service.

- ***V.xx Series Telephone Network Standards***

Table 17.4 summarized the *V.xx* series from the *ITU-T*. They include the most commonly used modem and telephone network standards. Prior to the *ITU-T* standards, the *American Telephone and Telegraph Company (AT&T)* and the *Bell System* offered standards such as Bell 103 and Bell 212A. Another set of standards, the *Microcom Networking Protocol (MNP)* Class 1 through Class 10 (there is no Class 8), but the development of an international set of standards will most likely prevail and continue to be extended. Typically, when modems "handshake" in communication, they agree on the highest standard transfer rate that both can achieve. Beginning with *V.22bis* (*bis* for "second version." *terbo* for "third version"), *ITU-T* transfer rates increase in 2400 Baud multiples.

Another industry standard, *ISDN* uses digitally encoded methods on phone lines to provide transfer rates up to 128,000 *Baud*. Another technology, *DSL*, provides even faster transfer rates.

ITU-T Standard	Meaning
V.22	Provides 1200 bits per second at 600 baud (state changes per second)
V.22bis	The first true world standard, it allows 2400 bits per second at 600 baud
V.32	Provides 4800 and 9600 bits per second at 2400 baud
V.32bis	Provides 14,400 bits per second or fallback to 12,000, 9600, 7200, and 4800 bits per second
V.32terbo	Provides 19,200 bits per second or fallback to 12,000, 9600, 7200, and 4800 bits per second; can operate at higher data rates with compression; was not a CCITT/ITU standard
V.34	Provides 28,800 bits per second or fallback to 24,000 and 19,200 bits per second and backwards compatility with V.32 and V.32bis
V.34bis	Provides up to 33,600 bits per second or fallback to 31,200 or V.34 transfer rates
V.35	The trunk interface between a network access device and a packet network at data rates greater than 19.2 Kbps. V.35 may use the bandwidths of several telephone circuits as a group. There are V.35 Gender Changers and Adapters.
V.42	Same transfer rate as V.32, V.32bis, and other standards but with better error correction and therefore more reliable
V.90	Provides up to 56,000 bits per second downstream (but in practice somewhat less). Derived from the x2 technology of 3Com (US Robotics) and Rockwell's K56flex technology.

Table 17.4

- **Modem**

Modem is an acronym for *Modulator/Demodulator*. This device connected to a computer, transmits and receives information data using a telephone line. *Modem* converts a digital data (in the computer at the sender end) into an analog data (telephone lines) then back to digital data (in the computer at the receiving end). *Modem* are manufactured either in external or internal format (*Figure 17.4*).

Internal Modem
a

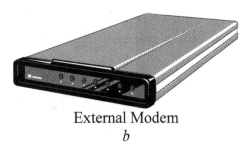

External Modem
b

Figure 17.4

Computer information is stored digitally, whereas information transmitted over telephone lines is transmitted in the form of analog waves. *RS-232* serial communication is a standard interface for connecting external modems (*Figure 17.4b*) to the computer. Consequently, any external modem can be attached to any computer that has an *RS-232* port, which almost all personal computers have. There are also modems that come as an expansion board (*Figure 17.4a*) that can be inserted into a vacant expansion slot. These are sometimes called *onboard* or *internal modems*. While the modem interfaces are standardized, a number of different protocols for formatting data to be transmitted over telephone lines exist. Some, like the *CCITT V.34*, are official standards, while others have been developed by private companies. Most modems have built-in support for the more common protocols at slow data transmission speeds, and can communicate with each other. At high transmission speeds, however, the protocols are less standardized. Aside from the transmission protocols that they support, the following characteristics distinguishing one modem from another:

o **Baud or bps:** How fast the modem can transmit and receive data. At slow rates, modems are measured in terms of baud rates. The slowest rate is 300 *Baud* (approx. 25 *cps*). At higher speeds, modems are measured in terms of *bits per second* (*bps*). 56K modems run at 57,600 *bps*, although they can achieve even higher data transfer rates by compressing the data. Obviously, the faster the transmission rate, the faster data can be sent or received. Note, however, that data cannot be received any faster than it is being sent. If, for example, the device sending data at 2,400 *bps*, the device receiving it would be at 2,400 *bps*. It does not always pay, therefore, to have a very fast modem. In addition, some telephone lines are unable to transmit data reliably at very high rates. 56K modems are better than 33K (33.6K *bps*) but often actually data rate is at 44K *bps* or so because of phone-line imperfections. Data may be received at that rate and not necessarily sent at the same speed rate.

o **Voice/Data:** Many modems support a switch to change between *voice* and *data modes*. In *data mode*, the modem acts like a regular modem. In *voice mode*, the modem acts like a regular telephone. Modems that support a *voice/data switch* have a built-in loudspeaker and microphone for voice communication.

o **Auto-Answer:** An *Auto-Answer modem* enables the computer to receive calls in the absence of the user.

o ***Data Compression:*** Some modems perform *data compression*, which enables them to send data at faster rates. However, the modem at the receiving end must be able to decompress the data using the same compression technique.

o ***Flash Memory:*** Some modems come with *flash memory* rather than conventional *ROM.* Built-in communications protocols can be easily updated if necessary.

o ***Fax Capability:*** Most modems are *fax modems*, i.e., they can send and receive faxes.

  • ***Cable Modems:*** Fast Internet access over upgraded cable TV networks. This uses existing cable TV infrastructure and a special modem. The downstream speed is from 384K *bps* to 30M *bps*. The upstream speed is about 300K *bps*.

  • ***Marrying Modems:*** Extra phone lines can be used to improve the Internet access using some specialized modems. Channel bonding or tandem modems can retrieve data from the Internet at twice or more the typical modem speed by taping two phone lines to do the downloading simultaneously. This method is close to the speed of most *ISDN* connections.

  • ***Multiplexing:*** *Multiplexing* is sending multiple signals or streams of information (data) on a carrier at the same time in the form of a single, complex signal, then recovering the separated signals at the receiving end.

  • ***Backbone:*** A *backbone* is a larger network transmission line that carries data gathered from smaller network lines that interconnect with it. At the local level, a *backbone* is a line or set of lines that *Local Area Network (LAN)* connect to for a *Wide Area Network (WAN)* connection or within a LAN to span distances efficiently, i.e. between buildings. At the Internet or other *WAN* level, a *backbone* is a set of paths that local or regional networks connect to for long-distance interconnections. The connection points are called network nodes or telecommunication *Data Switching Exchange (DSE)*.

  • ***Integrated Services Digital Network (ISDN):*** Introduced in 1988, the *ISDN* is a digital telephone service that operates using the standard copper phone wires but requires special adapters. It supplies two independent 64K *bps* channels that can be used for voice or data. *ISDN* is a set of *CCITT/ITU* standards for digital transmission over ordinary telephone copper wire as well as over other media. Basic *ISDN* data lines (transmit/receive) communicate at 56K *bps* but the maximum speed rate is at 128K *bps*. Home and business users who install *ISDN* adapters (in place of their modems) can see high quality graphic Web pages arriving very quickly (up to 128K *bps*). *ISDN* requires adapters at both ends of the transmission, i.e., the access provider also needs an *ISDN* adapter. *ISDN* in concept is the integration of both analog or voice data together with digital data over the same network. Although the *ISDN* installed is integrating these on a medium designed for analog transmission, *Broadband ISDN (BISDN)* will extend the integration of both services throughout the rest of the end-to-end path using fiber optic and radio media. *BISDN* will encompass frame relay service for high-speed data that can be sent in large bursts, the *Fiber Distributed-Data Interface (FDDI),* and the *Synchronous Optical Network (SONET).* *BISDN* will support transmission from 2 Mbps up to much higher rates. There are two levels of service:

1. *Basic Rate Interface (BRI),* intended for the home and small enterprises. *BRI* consists of two 64 Kbps *B channels* and one 16 Kbps *D channel*. Thus, a Basic Rate user can have up to 128 Kbps service.

2. *Primary Rate Interface (PRI)* is intended for larger users. *PRI* consists of 23 *B channels* and one 64 Kpbs *D channel* in the United States or 30 *B channels* and 1 *D channel* in Europe.

Both rates include a number of *B (bearer) channels* and a *D (delta) channel*. The *B channels* carry data, voice, and other services. The *D channel* carries control and signaling information.

- **Digital Subscriber Line (DSL):** *DSL* is a technology for bringing high-bandwidth information to homes and small businesses over ordinary copper telephone lines *(Plain Old Telephone Service – POTS)*. *DSL* deployment began in 1998. *xDSL* refers to different variations of *DSL*, such as *ADSL, HDSL, IDSL, RADSL, SDSL* and *VDSL*. In general, the maximum range for *DSL* without repeaters is 5.5 km (kilometers) or 18,000 ft. This is a point-to-point line, unlike cable which is shared in a given neighborhood or a geographic area. A home or a small business, that is close enough to a telephone company *Central Office (CO)* that offers *DSL* service, may soon be able to receive data at rates up to 6.1M *bps (Megabits per second)*. This would enable continuous transmission of motion video, audio, and even 3-D effects. More typically, individual connections will provide from 1.544M *bps* to 512K *bps* downstream and about 128K *bps* upstream. A *DSL* line can carry both data and voice signals and the data part of the line is continuously connected. *DSL* is expected to replace *ISDN* in many areas and to compete with the cable modem in bringing multimedia and 3-D to homes and small businesses. As the distance decreases toward the telephone company office, the data rate increases. Another factor is the gauge of the copper wire. The heavier 24 gauge wire carries the same data rate farther than the 26 gauge wire. *Compaq, Intel*, and *Microsoft (Universal ADSL Working Group – UAWG)* working with telephone companies have developed a standard and easier-to-install form of *ADSL* called *G.Lite* (also know as *DSL-Lite, G.992.2, Split-terless ADSL* or *Universal ADSL*) that is expected to accelerate deployment. *G.Lite* data rate is from 1.544M to 6M *bps* downstream and from 128K to 384K *bps* upstream. The *ITU* has standardized on three forms of *DSL: ADSL, HDSL* and *G.Lite*. The *G.Lite* standard is targeted toward the consumer and reduces the need for the phone company to install it at the home.

- **Asymmetric Digital Subscriber Line (ADSL):** *ADSL* is a data communication service that can deliver up to 8 *Mbps* downstream speed over an ordinary copper phone line. Until 1999, *ADSL* was in two options:

  1) *Carrier-less Amplitude/Phase modulation (CAP)* that uses a single carrier. Most *ADSL* connections in the *USA* use *CAP*.

  2) *Discrete Multi-Tone (DMT)* that uses multiple carriers. *DMT* is based on *ANSI* and the European standards for *ADSL*.

Each of these options was a different modulation system, which is a method of writing the data onto a carrier signal on the *ADSL* wire and for reading the data at the receiving end. *ADSL* data moves more quickly downstream (from the Internet) than upstream (to the Internet). *ADSL* uses special modem called *endpoint*. Along with this modem, the computer needs a *Network*

Interface Card (NIC) that treats the modem as a local device on the network. *ADSL* bypasses most of the typical *Public Switched Telephone Network (PSTN)*, the connection device that often delays the analog modems and *ISDN* connections with busy signals, ring-no-answer calls and other reliability issues. *ADSL* is a pure network service, connection is always ON; unlike the modem and *ISDN* adapters that must dial, connect and disconnect from the Net. *ADSL* line distance limit is 17,500 ft.

- **High-data-rate Digital Subscriber Line (HDSL):** *HDSL* is symmetric. It provides data speed of up to 2M *bps* in both directions (downstream & upstream).

- **ISDN Digital Subscriber Line (IDSL):** *IDSL* provides download data speed of up to 128K *bps*.

- **Rate-Adaptive Asymmetric Digital Subscriber Line (RADSL):** *RADSL* is a subset of *ADSL*, it automatically adjusts the data speed rate to the line condition.

- **Single-Line Digital Subscriber Line (SDSL):** *SDSL* provides download data speed of up to 6M *bps* but has a line distance limit of 10,000 ft.

- **Very-high-data-rate Digital Subscriber Line (VDSL):** *SDSL* is restricted to users within a mile from the phone company's *Central Office (CO)* but provides download data speed of up to 52.8M *bps*.

- **Digital Subscriber Line Access Multiplexer (DSLAM):** *DSLAM* is a network device, usually at the *CO*, that receives signals from multiple *DSL* connections and sends the signals on a high-speed *backbone* line using multiplexing techniques. Depending on the product, *DSLAM* connects *DSL* lines with some combination of *Asynchronous Transfer Mode (ATM)*, *frame relay*, or *IP* networks. *DSLAM* enables a phone company to offer business or home users the fastest phone line technology (*DSL*) with the fastest *backbone* network technology (*ATM*).

- **Satellite (DirecPC):** If the cable and *DSL* are not available, there is another modem connection, the *DirecPC from Hughes Networks*. The downstream speed is about 400K *bps*; this connection uses a small satellite dish pointed at the southern sky. This requires an analog modem for upstream transmission.

CHAPTER XVIII
ELECTRICAL NOISE & POWER

- *Basics Electrical Power*

As described in Chapter II, Electricity is the flow of electrons through a wire. These electrons provide the energy that drives, in this case, the computer and other devices. Electricity, for a PC, is described using four main characteristics:

o *Voltage:* The *voltage* on a line represents the force of the electricity. Voltage, also called *ElectroMotive Force (EMF)*, is an expression for electric potential or potential difference. If a conductive or a semiconductive path is provided between the two points having a relative potential difference, an electric current flows. The common symbol for voltage is the uppercase letter *V* or *E*. The standard unit is the volt, symbolized by *V*. One volt is the *EMF* required to drive one *coulomb* of electrical charge (6.24×10^{18} charge carriers) past a specific point in one second. Voltage can be either Direct or Alternating. A Direct voltage maintains the same polarity at all times. In an Alternating voltage, the polarity reverses direction periodically. The number of complete cycles per second is the *frequency*, which is measured in *hertz*. An example of pure direct voltage is the *EMF* between the terminals of an electrochemical cell. The output of a power-supply rectifier, prior to filtering, is an example of pulsating direct voltage. The voltage that appears at the terminals of common utility outlets is alternating. A potential difference produces an electrostatic field, even if no current flows. As the voltage increases between two points separated by a specific distance, the electrostatic field becomes more intense. As the separation increases between two points having a given potential difference, the electrostatic flux density diminishes in the region between them. A single charged object is surrounded by an electrostatic field whose intensity is directly proportional to the voltage of the object relative to other objects in its vicinity.

o *Current:* The *Current* is a measure of how much electricity is in the line; how many electrons are moving through in a given unit of time. Current is a flow of electrical charge carriers, usually electrons or electron-deficient atoms. The common symbol for current is the uppercase letter *I*. The standard unit is the *ampere*, symbolized by *A*. One *ampere* of current represents one *coulomb* of electrical charge (6.24×10^{18} charge carriers) moving past a specific point in one second. Physicists consider current to flow from relatively positive points to relatively negative points; this is called *conventional current* or *Franklin current*. *Electrons*, the most common charge carriers, are negatively charged. They flow from relatively negative points to relatively positive points. Electric current can be either direct or alternating. Direct current (DC) flows in the same direction at all points in time, although the instantaneous magnitude of the current might vary. In an alternating current (AC), the flow of charge carriers reverses direction periodically. The number of complete AC cycles per second is the frequency, which is measured in hertz. An example of pure DC is the current produced by an electrochemical cell. The output of a power-supply rectifier, prior to filtering, is an example of pulsating DC. The output of common utility outlets is AC. Current per unit cross-sectional area is known as *current density*. It is expressed in amperes per square meter, amperes per square centimeter, or amperes per square millimeter. Current density can also be expressed in amperes per circular mil. In general, the greater the current in a conductor, the higher the current density. However, in some situations, current density varies in different parts of an electrical conductor. A classic example is the so-called *skin effect*, in which current density is high near the outer surface of a conductor, and low near the center. This effect occurs with alternating currents at high frequencies. Another example is the current inside an active electronic component such as a *Field-Effect Transistor (FET)*. An electric current always produces a magnetic field. The stronger the current, the more intense the magnetic field. A pulsating DC, or an AC, characteristically produces an electromagnetic (EM) field. This is the principle by which wireless signal propagation occurs

o **Volt-Amps:** This is the product of *voltage* and *current*. It represents the total "raw" amount of power being supplied by the electrical source. It is measured in *volt-amps,* and its symbol is *VA*.

o **Power:** The *Power* is the amount of actual electrical work being performed over a period of time. Electrical power is the rate at which electrical energy is converted to another form, such as motion, heat, or an electromagnetic field. The common symbol for power is the uppercase letter *P*. The standard unit is the *watt*, symbolized by *W*. In utility circuits, the kilowatt (kW) is often specified instead; 1 kW = 1000 W. One watt is the power resulting from an energy dissipation, conversion, or storage process equivalent to one joule per second. When expressed in watts, power is sometimes called *wattage*. The wattage in a direct-current (DC) circuit is equal to the product of the voltage in volts and the current in amperes. This rule also holds for low-frequency alternating-current (AC) circuits in which energy is neither stored nor released. At high AC frequencies, in which energy is stored and released (as well as dissipated or converted), the expression for power is more complex. In a DC circuit, a source of E volts, delivering I amperes, produces P watts according to the formula:

$$P = E * I \qquad (18.0)$$

When a current of I amperes passes through a resistance of R ohms, then the power in watts dissipated or converted by that component is given by:

$$P = I^2 * R \qquad (18.1)$$

When a potential difference of E volts appears across a component having a resistance of R ohms, then the power in watts dissipated or converted by that component is given by:

$$P = \frac{E^2}{R} \qquad (18.2)$$

In a DC circuit, power is a scalar (one-dimensional) quantity. In the general AC case, the determination of power requires two dimensions, because AC power is a *vector quantity*. Assuming there is no reactance (opposition to AC but not to DC) in an AC circuit, the power can be calculated according to the above formulas for DC, using *root-mean-square (rms)* values for the alternating current and voltage. If reactance exists, some power is alternately stored and released by the system. This is called *apparent power* or *reactive power*. The resistance dissipates power as heat or converts it to some other tangible form; this is called *true power*. The vector combination of reactance and resistance is known as impedance.

If the load on the electrical line is a light bulb, *Power* is equal to *volt-amps*. The amount of power consumed is equal to the amount of power supplied. The PC power supply, however, is a complex load, because of the way that it works to convert the power from the power line to the type used inside the PC. In this case, converting between volt-amps and wattage requires a conversion factor. Power equipment manufacturer typically use a conversion factor of 1.4, or

$$volt\text{-}amps \text{ of a } PC = 1.4 * Power \qquad (18.3)$$

The *volt-amps* and *Power* are important to the computer user. The amount of power used by the computer and devices is typically expressed in watts, while the power supplied by *Uninterruptible Power Supply (UPS)*, for example, is usually expressed in *volt-amps*. This conversion factor help calculate the *volt-amps* if a *UPS* is needed.

- **The Power Supply**

The internal power supply is responsible for converting the standard household electricity into a form that the computer can use. The power supply is responsible for powering every device in the computer; if it has a problem or is of low quality. The PC may experience many difficulties that may not be realized by the user and that was actually the fault of the electrical system which is the power source to the power supply. The power supply plays an important role in the following areas of the system:

- o **Stability:** A high quality power supply with sufficient capacity to meet the demands of the computer will provide years of stable power for the PC. A poor quality or overloaded power supply will cause electrical glitches that are particularly insidious, because the problems occur in other, seemingly unrelated, parts of the system. For example, power supplies can cause hard disks to develop bad sectors or cause software bugs to appear, these problems can be very difficult to trace back to the power supply.

- o **Cooling:** The power supply contains the main fan that controls the flow of air through the PC case. This fan is a major component in the PC's cooling system.

- o **Energy Efficiency:** New PC power supplies work with the computer's components and software to reduce the amount of power they consume when idle. This can lead to significant savings over old PC systems.

o **Expandability:** The capacity of the power supply is one factor that will determine the ability to add new devices to the system, or upgrade to a more powerful motherboard or processor. Many people do not realize, for example, that a Pentium Pro processor and motherboard consume far more power than a similar 486-based system, and the power supply needs to be able to provide this power.

- **Power Supply Functions and Signals**

The power supply's main function is simple: get the input from the power source system of the home or office, and convert it into a power that the PC can use. However, the PC power supply's job is not as simple as that of a standard power converter. The PC power supply must provide several different voltages, at different strengths, and must also manage some additional signals that the motherboard uses.

o **AC-DC Voltage Conversion:** Standard household current is supplied as an alternating current (AC) of 110 volts (in North America) or 220 volts (Europe and Middle East). Alternating current is so named because the signal actually alternates between positive and negative 60 or so times per second (sine wave). This type of power is suitable for many types of devices, but totally unsuited for use within low-voltage devices such as computers. The power supply in the system converts this line power into direct current (DC), which is what the computer requires for operation. DC power is the sort of power that would get from a battery--it stays at a constant level. The external peripherals, however, such as printers, external modems, external storage drives, may have an "AC adapter" which looks like a little, heavy black box that plugs into the wall outlet (some may have a thin wire that comes out of the black box and into the peripheral). That little black box is also a power converter, changing the AC power from the wall into DC power that the device uses. Usually, the input and output electrical specifications are printed right on it.

o **Voltages Source:** A PC power supply provides the following voltages to the motherboard and drives:

- **+5V DC:** This is the standard voltage that is used by the motherboard and most of the circuitry on peripheral cards in the computer. Wires carrying +5V power are normally RED in color.

- **+12V DC:** This voltage is mostly used to power disk drive motors and similar devices. It is in most cases not used by the motherboard in a modern PC but is passed on to the ISA system bus slots for any cards that might need it. Wires carrying +12V power are normally YELLOW in color (but are also sometimes red).

- **-5V DC and -12V DC:** These were once used on older systems and are included for compatibility reasons, but are not generally used in new machines. The output specifications printed on most power supply, these voltages are supplied with very low current (less than 1 amp normally) because they are not usually used for much of anything at all.

- **+3.3V DC:** Older processors ran at the same +5V that most of the motherboard runs at. However, most 486 class chips that are 100 MHz or above, all Pentiums that are 75 MHz or above, and all Pentium Pro, Pentium II or equivalent chips, are 3.3V chips (some new CPUs use even lower voltages internally). Old boards must further convert the 5V signal from the power supply into the 3.3V that the processor needs. This requires a voltage regulator on the motherboard, which adds to its cost and generates a substantial amount of heat. The newer style ATX power supply can provide the 3.3V power for the CPU directly, which is more efficient and reduces heat and cost for the motherboard.

- **Processor Power and Voltage:** In the early days of computers, there was not very much concern about how much power a processor used. There were not as many of them, and not nearly as much to do with them. As time has gone on the users' demands, these machines have continued to increase and new uses have put power consumption in the spotlight. This has led to a confusing set of voltage specifications where before (up to the Intel 486DX2-66) everything ran on 5 volt power. The power usage and the voltage support of a processor are important for the following reasons:

  - Power consumption equates largely with heat generation, which is a primary enemy in achieving increased performance. New processors are larger and faster, and keeping them cool can be a major concern.

  - With millions of PCs in use, and sometimes thousands located in the same company, the desire to conserve energy has grown from a non-issue to a real issue in the last five years.

  - Reducing power usage is a primary objective for the designers of notebook computers, since they run on batteries with a limited life. They are more sensitive to heat problems since their components are crammed into such a small space.

New processors strive to add additional features and to run at faster speed, which tends to increase power consumption. Processor designers compensate for this largely through technology, by using lower-power semiconductor processes, and shrinking the circuit size and die size.

- **Power Good Signal:** When the power supply first starts up, it takes some time for the components to get "up to speed" and start generating the proper DC voltages that the computer needs to operate. Before this time, if the computer were allowed to try to boot up, strange results could occur since the power might not be at the right voltage. It can take a half-second or longer for the power to stabilize, and this is an eternity to a processor that can run 200 million instructions per second! To prevent the computer from starting up prematurely, the power supply puts out a signal to the motherboard called *"Power Good"* after the power is ready for use. Until this signal is sent, the motherboard will refuse to start up the computer. In addition, the power supply will turn OFF the *Power Good* signal if a power surge or glitch causes it to malfunction. It will then turn the signal back ON when the power is OK again, which will reset the computer. If there is a *brownout*, where the lights flicker OFF and ON for a split-second, the computer, however, seems to keep running but resets itself, that is probably what happened. Sometimes a power supply may shut down and seem "blown" after a power problem but will reset itself if the power is turned OFF for 15 seconds and then turned back ON.

• ***Power ON and 5V Standby Signals:*** In addition to 3.3V DC power for the processor, the new *ATX* style power supply also adds functionality to permit software control of the power supply, for motherboards and software that support it. The "Power On" signal can be controlled by the motherboard to tell the power supply when to turn itself OFF. The "5V Standby" is a special 5V DC source to the motherboard separate from the normal 5V. It remains on even when the rest of the system is powered down, which lets special circuitry in the motherboard function even when the normal power is OFF. This could be used, for example, to allow a software re-start of the computer even from a totally off condition.

• ***Power Supply Loading:*** A PC power supply will only function properly if it has a load to drive. Because of the method used to generate the DC voltage (called *switching*), the supply must have something to draw at least some power or it will not function at all. This is why a system without a hard drive of some sort in it will normally fail to function properly: there is nothing to draw power from the +12V line. (The fan built in to the power supply does draw some of this 12V power but typically not very much). It is recommended not to turn ON the power source (wall outlet) to the power supply of the PC to "test it" without having a load, such as a motherboard, a hard drive or any other device, connected to the PC power supply. Without a load, it may result in either the power supply shutting itself down, or in the case of some low-cost units, damage or destruction of the power supply itself. New and better power supplies have special resistors built in to prevent this from happening, but it is not worth the risk, particularly since there is really no reason to even try doing it.

• ***Power Switch:*** The old style PC case had the power switch at the back of the machine, usually on the right side of the case. This switch was actually inside the power supply itself, with a hole cut out in the case so that it could be reached from the outside. New PC machines move the power switch to the front of the case. *AT form factor* systems use a remote, physical toggle power switch that is connected to the power supply using a four-wire cable (sometimes with a fifth wire, a grounding line). The switch is normally mounted to the front of the case. Some "slimline" cases actually use a mechanical plastic stick that is pushed on by the button on the front of the case and presses against the real switch on the power supply itself, in the back of the machine. **Warning:** The remote power switch wires on an AT system carry live 110V (or 220V) AC power whenever the power supply is plugged in, even when the power is OFF. Safety first, always unplug the power from the wall outlet, even if the PC is OFF, AC power may still be at the switch. The new *ATX form factor* changes how the power switch works altogether. Instead of using a physical toggle switch connected to the power supply, on ATX systems the power switch is electronic. It connects to the motherboard, much the way that the reset switch does. On an ATX system, when the power switch is pressed, a "request" is sent to the motherboard to turn the system ON. This design has several advantages over the older AT style power switch, including giving the PC the ability to turn itself ON and OFF under software control, if the appropriate supporting hardware and software are used.

• ***No Protection:*** It is highly recommended to have a power protection. If cost is a major concern, or if the internal power supply is of superior quality, one electrical source strike could damage part if not all the computer. Always be safe and careful, the hardware can be replaced in minutes or hours, PC data and information may take days or weeks to recover.

- ***Surge Suppressors:*** Most people at the very least use a surge suppressor to protect their PC system. This usually takes the form of a plastic block into which the plug in the computer, monitor and other devices, with a cord which then plugs into the wall. These are also sometimes called "*power strips*" or "*power outlets*". Surge protectors range in quality from very good to almost useless, with the protection being somewhat proportional to the cost. Low-cost surge suppressors do not really offer very much protection. Surge suppressors normally work to reduce power problems in two ways. First, they use power absorbing components that take the shock of voltage spikes and surges and prevent them from being passed on to the computer equipment. Second, they normally include some line conditioning circuitry to smooth out and reduce power line noise, although not the top of the line components but a true line conditioner. Not all surge suppressors condition the line. Some of the quality and feature considerations that should be taken into account when shopping for a surge suppressor:

    o ***UL Standards for Surge Suppression:*** Underwriters Laboratories established standard UL 1449 for surge suppressors. Any suppressor meeting this standard is probably one that will provide useful protection. There are three levels of protection: 330, 400 and 500. This number refers to the maximum voltage that the suppressor will allow to pass through the line; the lower the number, the better.

    o ***Energy Absorption:*** A surge suppressor also has an energy absorption or dissipation rating, measured in Joules. The higher the value, the more energy the suppressor can absorb, so higher is better. Surge suppressor 200 is basic protection, 400 is good protection and 600 is superior protection.

    o ***Line Conditioning:*** Surge suppressors normally contain some line conditioning capabilities as well.

    o ***Protection Indicators:*** Many surge suppressors now come with an LED that is supposed to be lit when the line is protected. With most good suppressors, this indicator can be relied on, but with low-cost ones, it often cannot. In some cases, over time a suppressor can absorb enough power to degrade the components' ability to protect the PC, but not so much in one direct hit that the LED will be turned OFF. If the LED is out (not working), the suppressor needs to be repaired or replaced. If there is no LED, it is difficult to know the status of the surge suppressor.

      Note: Some suppressors have a light that is just a "POWER ON" light. This obviously provides no indication at all of the protection level, but it does tell if the suppressor is plugged in and working (at least at a basic level).

- ***Line Conditioners:*** Line conditioners work by filtering and smoothing the power stream to eliminate dips, fluctuations and interference that can cause power to be "noisy". Their ability to reduce noise is measured in decibels over a given frequency range (just like noise reduction in the home stereo system). The more noise reduction, the better. True line conditioners can be quite costly due to the high quality of components required to do the best job. Many other power protection devices (such as surge suppressors) incorporate simple line conditioning circuitry, at a lower price. In practice, few PC users employ line conditioners.

- ***Protecting the Modem and Other Peripherals:*** Voltage spikes can be carried along any convenient wire and into the PC. In particular, wires that run between buildings are susceptible to major disruptions due to lightning--which is why it is illegal in many places to run copper network wiring outdoors. After the power line, the next biggest problem area is the telephone line. Lightning can be carried along the line and into the home, damaging the modem and possibly even the motherboard or other components. It is also possible for a spike to be carried along a networking cable, causing similar effects. In fact, a spike on a network line can damage every PC on the network. Most good power protection systems include protection for the modem line. If a modem is being used in a PC then a system that will protect it from any electrical disturbance should be used as well. The network will generally be protected if every PC on it is properly protected, as long as there is not any network cables outside between buildings.

- ***External Power Problems:*** There are several types of problems that the computer's internal power supply will be subjected to if the electrical power is from a wall electrical outlet. Some minor problems the computer system may be able to deal with, while others can cause data loss or even permanent damage to the PC. Problems with the PC system power should not be underestimated as a potential source of virtually any hardware-related PC troubles such as,

  o ***Line Noise:*** Line noise consists of small variations in the voltage level delivered to the computer. A certain amount of line noise is normal (no power generation circuit is perfect). In some areas, however, the power quality is worse than others. Also, if the PC is sharing a circuit or is physically located near devices that cause electromagnetic interference (motors, heavy machinery, radio transmitters, etc.) then electrical line noise can be a concern. Electrical noise that the power supply cannot handle can cause it to malfunction and pass the problem on to the motherboard or other internal devices. Some devices and appliances in the home or the office can produce line noise and pass it on to other devices through the power system. Better power protection devices will isolate devices that plug into them to prevent this cross-pollution. *Figure 18.0* shows a DC pulse with an electrical noise,

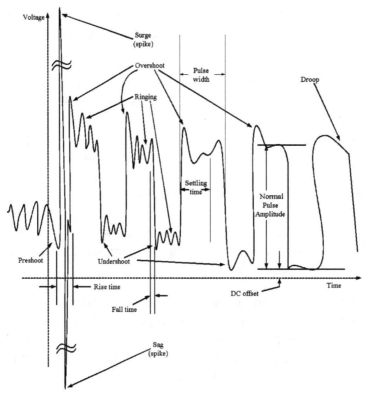

Figure 18.0

Figure 18.1 shows a distorted DC pulse without an electrical noise,

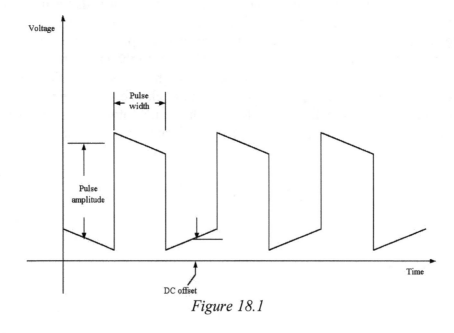

Figure 18.1

Figure 18.2 shows an ideal DC pulse without an electrical noise,

212

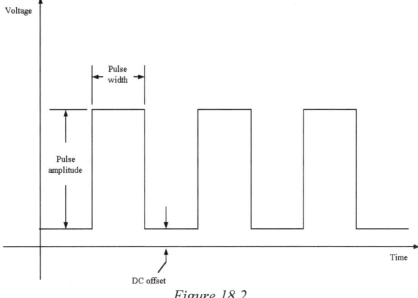

Figure 18.2

o **Surges:** The electrical power from the wall outlet is rated by the electrical company to be within a certain voltage range. The nominal voltage for North American circuits is 110 *V* AC (220 *V* AC in Europe and the Middle East). Due to disturbances, distant lightning strikes, and problems within the electrical grid, on occasion a voltage spike may come down the line. This is a temporary increase of voltage that can last just a few thousandths of a second, but in this time the voltage can increase from 110 to 1,000 volts or even higher. Most computer power supplies are subjected to many of these surges each year. High voltage surges can disrupt or even damage the computer equipment. In addition, being subjected to many surges over a period of time will slowly degrade many power supply units and cause them to fail prematurely.

o **Lightning Strikes:** Lightning can sometimes deliver a charge of millions of volts for a few micro or milliseconds. If the home is hit directly, significant damage will likely be the result. Good line protection equipment can often help reduce the chances of catastrophe should lightning strike. A strike near the home will likely result in a voltage spike on the line, normally a very strong one, enough to do severe damage to a PC.

o **Brownout:** A *brownout*, or *voltage sag*, is a "dip" in the voltage level of the electrical line. When a brownout occurs, the voltage drops from its normal level to a lower voltage and then returns; in some ways, it is the opposite of a surge. It is extremely common, and can lead to mysterious problems that cannot be blamed on the power system. A brownout can sometimes detected by noticing the lights flickering or dimming; this occurs often during heavy load periods such as in the late afternoon on a hot summer day. Brownout can damage devices within the computer systems. In many ways, they are worse than a blackout. In a blackout, the power just goes OFF, but with a brownout the device continues to get power but at a reduced level, and many devices will malfunction rather than fail totally.

213

o **Blackout:** A *blackout*, or *disruption*, is when the power totally fails. The damage that a blackout causes to the computer system depends a great deal on its timing. If the system is idle when the power goes out, probably nothing will be damaged within the system when the power comes back ON. However, if the power goes OFF while using the PC, fixing hardware problems on the PC or just as the drive was updating the FAT, very likely the PC will have a problem or more.

In addition, the power usually does not go out cleanly, there will be electrical spikes and jitters seen as the power is going OFF and then comes back ON. Most systems survive the power going OFF and back on without difficulties, but there is the potential for large amounts of damage.

- *Protection Against Power Problems*

There are several levels of protection that can be applied to help protect the computer system, the applications and the data from any loss or damage as a result of these power problems. It is not always true that the more money is spent on power protection devices, the better PC protection will be. In addition, using an *Uninterruptible Power Supply (UPS)* will allow the PC to avoid most power problems with protection against power service interruptions.

- *Uninterruptible Power Supplies (UPS)*

The *UPS* is a device with special circuitry and batteries that are used to prevent the PC from losing power during a *blackout* (*disruption*) or a *brownout* (*voltage sag*). *UPS* generally comes in two types: *On-line* or *Active UPS*.

o **Standby UPS:** A *Standby UPS* is an uninterruptible power supply that provides continuous power by switching from standard AC power to a backup battery in the event of an interruption. Most of the time the machine runs OFF the normal power line. When the power is disrupted, the *UPS* instantly switches over to the battery, and when it is restored the battery is switched back. The *Standby UPS* usually functions well, but there *is* a switching time to take into account. If the *UPS*'s switching time is too slow, or the computer is very sensitive, the PC might reboot when the switch takes place, which would defeats the purpose of a *UPS*. In most cases, this will not happen, but a *Standby UPS* may not be the perfect power protection for an important server system. If a *Standby UPS* is used, it is important to incorporate surge suppression and line conditioning features for when the machine is running off standard power, this is a protection from noise and power disturbance.

o **On-Line UPS:** An *On-Line, active* or "*true*" *UPS* functions differently from a *Standby UPS*. With an *On-Line UPS*, instead of switching to a battery when needed, the system always runs off the battery. Two voltage converters run continuously and independently inside the *UPS*: one converts the battery power (DC voltage) into AC voltage for the system to use, and another converts power from the wall outlet to recharge the *UPS*. If the power goes out, only the recharging the battery stops. The system keeps running off the battery, for as long as it lasts. An *On-Line UPS* is the best power protection available. Since there is no battery switchover, there is no chance of power interruption when the power goes out. The computer is never running directly off the AC line, and this isolation means *spikes*, *line noise* and *brownouts* are virtually eliminated. The cost of the *On-Line UPS* is higher than the *Standby UPS*.

- ***Battery Power and Battery Service Time***

  The more power the PC draws, the more powerful a *UPS* is required. When choosing a *UPS*, enough power to drive the PC system will be needed. If not, the *UPS* may fail to keep power to the PC when needed. In general, a lower-powered unit will give less run time before it fails compared to a higher-powered one. However, a unit that is vastly underpowered for the machine it is driving may fail immediately, with zero run time, thus making it useless. UPS power is normally expressed in terms of *VA*, or *volt-amps*. *UPS* manufacturers often recommend a conversion factor of 1 *VA* = 1.4 *Watts* for normal PC use. For example, a 350 *VA UPS* would be sufficient for a 250 *W* PC power supply. The 350 *VA UPS* will power only a minimal computer system, i.e. that does not include the monitor or any external devices (tape backup, scanner, etc.). That also needs to be calculated. The wattage required for the monitor and other devices may be significant and is not part of the 250 W that the PC power supply is rated for. Figuring out how much *UPS* is needed can be difficult. Some suggestions may help facilitate the decision:

  o ***Read the Recommendations:*** Look at the recommendations on the box of a *UPS* under consideration. In general, a 350 *VA UPS* is only going to cut it for a low-end PC. The more devices installed in the PC, the more power it is going to use.

  o ***Use UPS manufacturer Sizing Tool:*** Some *UPS* manufacturer may have a UPS sizing tool, use it for an estimate math.

  o ***Test the UPS:*** An obvious way to see if the *UPS* meets the PC needs: test it. While the system is running, pull the plug on the *UPS*, and see how long the power lasts before the *UPS* shuts down. If the *UPS* works properly and allows time enough to close and save PC work then shut down, protection has been established.

  o ***Warranty and support:*** Check for the UPS warranty and the support is offered.

- ***Electrical Waveform Output***

  A normal *AC* electrical signal is a relatively smooth sine wave, and a good *UPS* will produce a sine wave as its output to approximate this. Some low-cost *UPS* will produce a square wave instead (*Figure 18.4*). With some power supplies, this square wave can cause problems, because the supply is expecting the sine wave. Better supplies will handle either waveform, but sine wave output is still preferred.

- ***Status Indicators***

  Most *UPS* units come with some a status indicator. This may be something a light that indicates the unit is functioning or a readout that shows when the unit is running off battery power, (or even how much time is left on the battery power).

- ***Battery Replacement***

  Like all batteries, eventually, the *UPS* battery will have to be replaced. How quickly they have to be changed depends on the unit and how often it is put in use. Some units have user-replaceable batteries, while others require the UPS to be sent back to the factory for replacement. Obviously, user-replaceable battery is far more convenient.

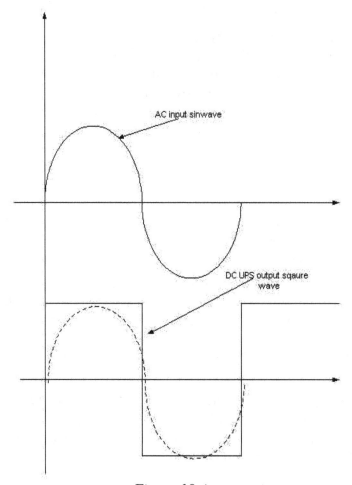

Figure 18.4

- **Automatic Shutdown**

If the power goes out when the PC is ON and idled, then the *UPS* cannot be of any help to the system unless there is some way to have PC *"gracefully"* shut down automatically. Fortunately, some *UPS* have a cable that connects to the PC's serial port and allows the *UPS* to send a signal to the PC when the line power goes out. Software running on the PC performs a shut down of the system so that no data file corruption occurs when the power goes out. New systems utilizing the ATX style motherboard and power supply actually have the ability to shut the machine down completely, using a signal that can be sent to the power supply from the motherboard.

- **Comparison of Power Protection Methods**

Table 18.0 compares the different power protection options, showing the approximate amount of protection each offers against common power problems. The protection levels shown are typical, but depend very much on the individual device. The values for the *Standby UPS* assume that it also incorporates a surge suppressor, since most decent quality ones do. If it does not, then the protection level for surges and lightning strikes will drop to the same protection levels for those problem types as listed under *"No Protection"*.

216

Protection Type	Line Noise	Surges	Lightning Strikes	Brownout	Blackout	Cost
No Protection	Low to High	Low	Very Low	Very Low	None	None
Surge Suppressor	High	High	Low to Med	Very Low	None	**Low**
Line Conditioner	**Almost Perfect**	Very High	Med to High	Low to Med	None	High
Standby UPS	Med to High	High	Med to High	Almost Perfect	Very High	Med to High
On-line UPS	**Almost Perfect**	**Almost Perfect**	**High**	**Perfect**	**Almost Perfect**	Very High

Table 18.0

CHAPTER XIX
COMPUTER OPERATION

In reality, the computer consists of an application running on top of three essential layers: the *Operating System*, the *BIOS*, and the *Device Driver*. All these three together make the computer come active. In computer language, a command typed from a keyboard or read from a program instructs the CPU what to do. A command is a sequence of letters or numbers that directs the action of the computer.

A computer program is a sequence of instructions written in a computer language. The set of instructions tells a computer step-by-step exactly how to execute a task. A task is the software instruction that forms the lowest unit of a process. The size of the CPU instruction set directly affects the most fundamental issues involved in a computer design. A small and simple instruction set offers advantage and simplicity of the hardware design, but increases the program size. A large and complex instruction set decreases the program size, but increases the hardware complexity. To address these two types of computers, the *RISC* and *CISC* computers were developed.

In the creation of a computer program, there are four major areas that need to be considered:

    a) Algorithms and Data Structures.
    b) Architecture.
    c) Programming Languages.
    d) Software Methodology.

There are many levels of computer language, but all software programs follow the same path to the CPU, as shown in *Table 19.0*,

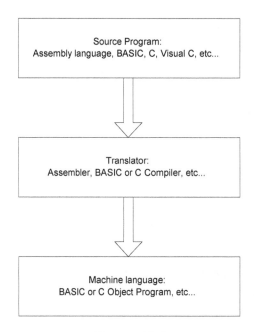

Figure 19.0

Soon after the introduction of the *ENIAC,* computers were programmed primarily in *Machine Language*. In the early 1950's, symbolic machine language called *Assembly Language* was introduced to make it easier to program the computer. In the late 1950's and early 1960's, the first high-level language was introduced: *Formula Translator* or *FORTRAN* by *John Backus*. During the 1960's, *Grace Murray Hopper* (*Figure 19.1*) developed the *Common Business Oriented Language* or *COBOL*. In the 1970's, *Niklaus Wirth* developed *PASCAL*. Until the introduction of the terminal (attached to a mainframe computer), most programming was entered into the computer through a punched card (*Figure 19.2*).

Grace Murray Hopper
Figure 19.1

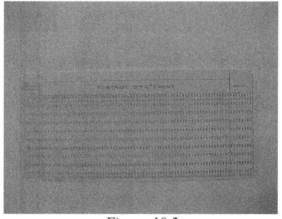

Figure 19.2

In every process, there is an order of data flow. Data must be manipulated or processed. For the computer, this data needs to be changed into information that it can understand.

As described in Chapter I (*Figure 1.13*), the data flow has three steps but a fourth step is essential, (*Figure 19.3*).

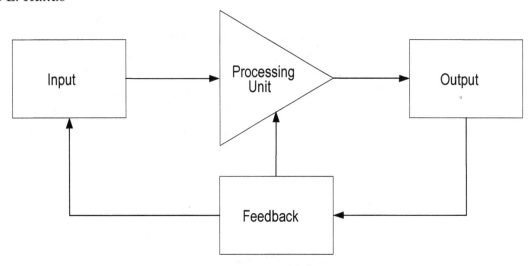

Figure 19.3

Input Unit: The *Input Unit* in a computer system is the process of capturing data and organizing it in a form that the computer can "understand". This process could include both the data that is manipulated and the software to do it.

Input can be by typing characters from a keyboard, using a scanner, speaking into a microphone, or reading a file from a floppy disk. In other words, *Input* involves three steps:

  a) *Collecting "raw" data and assembling it into a location.*
  b) *Checking and verifying the completeness and accuracy of this data.*
  c) *Converting and coding this data into a machine language that is readable by the CPU to be processed.*

Processing Unit: Once the data is captured into a computer, it is processed. Processing this data occurs in the CPU through logical and arithmetic operations with help from many devices such as the *Primary Memory* (RAM) and *Secondary Memory* (storage devices: hard disk, etc...). While processing, data may be accessed and re-accessed from memory, written and re-written into memory. This processing of data may entail many types of manipulation, such as:

  a) *In order for the data to be meaningful to the user, it needs to be classified and categorized according to certain characteristics or groups: personnel, inventory, cars, etc...*
  b) *The data should be sorted and arranged (alphabetically or numerically): last names, rank number, etc...*
  c) *The data is calculated arithmetically or logically, a result or a decision must be concluded: GPA of students, open employee gate, etc...*
  d) *The data is summarized, e.g., reducing it to a concise and usable format: Sales figures of the month by a salesperson, Dean's list based on GPA summary, etc...*
  e) *The data is now stored and retained: saved on a floppy or hard disk for later retrieval.*

Output Unit: the computer processes the data, the conclusion can now be distributed in different types of output:

  a) **Soft:** It is information that can be viewed on a monitor (screen). This is temporary because when the computer is shut down and the monitor is OFF, the information is gone.

b) ***Hard:*** It is the information that is printed in a tangible form on paper. This information can be read away from the computer (ON or OFF).

The data flow in the output phase consists of these steps:

   a) *Data is retrieved from memory (RAM or storage).*
   b) *Data is converted into a form that a user can understand and use: words, numbers, pictures, etc...*
   c) *Data is communicated back. It provides the necessary information to the selected users at the proper time and in an appropriate format.*

Feedback Unit: The processed data or information is monitored and evaluated at the output phase. Feedback is this process where now adjustments can be made to the input or the method it was processed. This will ensure that the processing continues to produce an output of satisfying information to the user.

- ***Operating System (OS)***

The *Operating System* is the soul of the computer. It controls the hardware through the instructions of the software: The display of graphics and text on the monitor, the access (Read and Write) of hard drive, floppy drive, CD ROM, tape backup, etc... the sound of the speakers, the printer output and so on. The Operating System is a software program that controls the operation of a processor system by providing input/output signals, allocation of memory space or translation of computer programs. It is an organized collection of procedures and programs for operating the computer. These programs interface users to computers and allow them to interact with the system in an efficient manner.

- ***Basic Input/Output System (BIOS)***

The BIOS is a special program of software stored in most computer memory. It is also known as the firmware and is the closest software language that drives the hardware circuitry of the computer. It is also the heart and soul of the computer, it controls the startup process of the computer. The firmware functions as a bridge between the CPU of the computer and the Operating System. BIOS software program is stored in ROM or Flash ROM.

- ***Device Drivers***

In any computer, there are many specialized software programs, which are like BIOS. These programs are specific to a hardware component in a computer such as Video card, CD ROM, Sound card, scanner, etc... The purpose of the device driver is to communicate with the Operating System and the BIOS of the computer for the specific component.

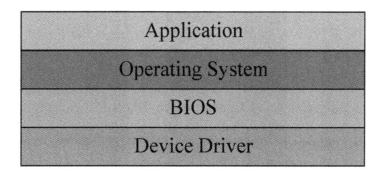

Figure 19.4

Figure 19.4 shows an example of the cooperation between the Operating System, BIOS and Device Driver: A document is to be stored on a floppy disk. The user commands the application software (word, spreadsheet, database, etc...) to save the document, a command is sent from the application software to the Operating System. The Operating System now checks the command of the user by verifying that there are no problems with the data and the name of the document, the destination drive (if it exists on this computer). This command is then sent to the BIOS for the selection of the destination (a floppy disk). The command now is sent to the Device Driver, which translate the software instructions into electrical signals that begin to spin the disk inside the floppy drive and write the data of the document onto the floppy disk. The process is then reversed to verify that the document has been saved successfully (i.e., the Operating System now reads the data from the disk and compares it with the document of the application).

- ### *Disk Operating System (DOS)*

In 1981, when the Personal Computer was first introduced to the market, there was no hard drive, but instead a Single Sided floppy drive. In order for the computer to start, it needed an Operating System software program. *Microsoft* introduced a *Disk Operating System* (*DOS*) floppy disk to start the computer. Once the computer is boot-up, an application software (such as word or spreadsheet program) disk can run. Shortly after the floppy drive introduction, the hard drive was introduced; DOS program was then copied or installed onto the hard drive.

- ### *Microsoft Windows*

In 1983, *Microsoft* introduced the Interface Manager (later called Windows) software to add some visual commands and graphical pictures to DOS. Windows was not very popular until 1990 with version 3.0; DOS was required to boot-up the computer. In 1993, Microsoft then introduced the *Windows NT (New Technology)* as a powerful Operating System for desktop computers as well as server systems. In 1995, Microsoft introduced the Windows 95 Operating System that would no longer require DOS boot sector. In 1996, Microsoft introduced Windows CE for handheld PCs, TV set-top boxes and other computer systems beyond the regular desktop PCs. In 1998, Microsoft introduced Windows 98, the successor to Windows 95.

- ### *The computer at the "BOOT" stage*

Whenever a computer is turned ON, there is a process of operation that it goes through: The **cold boot** or **boot-up** where the computer is bringing all its components to active by loading the **O**perating **S**ystem software. A computer cannot run without an **Operating System (OS)**. An *OS* is a software program with instructions that turn ON and OFF electrical circuits' lines. These lines make up the logical path of the computer. This software sets the protocols, rules and process for using electronic integrated circuitry such as memory (RAM, ROM, EPROM, etc...), drives (floppy, hard, removable, etc.), cards (video, sound, SCSI, etc...) and other components inside the computer.

During the boot, the *CPU* checks of all components, this is the *Power On Self-Test (POST)* program. The *CPU* gets all its instructions on what to do from a software program that is stored in a memory ROM. This is the *BIOS* program. When the switch of the computer is ON, the *CPU* first checks itself and the *POST* program. A signal, initiated by the *CPU* is sent over the BUS, all components are electrically connected to each other. The *CPU* can verify the function of all components through a programmed routine. The *CPU* then checks the system's clock; this is a vital function for the computer, it is the heartbeat of the computer. The clock is responsible for keeping the computer's components synchronized when electrically communicating to each other. An electrical pulse signal clears the memory ICs (data) by resetting the program counter and memory registers. The *POST* then checks the memory *RAM* by running a series of tests. These tests, which are preprogrammed routines in the *BIOS*, are to ensure that the memory chips are functioning properly in write and read mode. The *POST* checks the attachment status of the keyboard (also checks if any key has been pressed) and the pointing device (mouse).

If the *CPU* detects an error (memory, keyboard, plug-in card, or other essential components), it normally results in a beep with an error message on the monitor (display) or a sequence of beeps sound. The number of beeps separated by a pause can be translated into the specific problem of the computer. If the *POST* passes and the *CPU* does not detect any error, a single beep combined with a text or graphics on the monitor (display). At this time, the *CPU* reads the next sets of instructions from the system files, which are stored on the hard drive. A healthy and live computer has a healthy and live firmware: an ideal marriage of a software language to a hardware circuit.

A *soft* or *warm boot* is a restart (reset) of a running (working) computer by pressing the combination keys of *CTRL*, *ALT* and *DEL* all at once from the keyboard of the computer.

During the *POST* programs, which are performed each time the computer system is powered ON, errors may occur. Following the flow chart in *Figure 19.5*, there could be three possibilities:

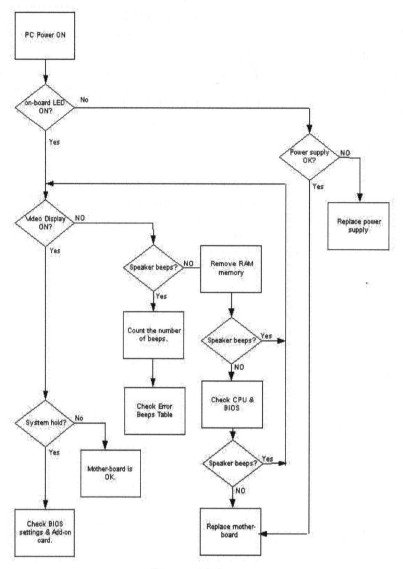

Figure 19.5

a) *No errors*: Computer system booted-up and ready to be used.
b) *Non-fatal errors*: These errors, in most cases, allow the computer to continue the boot-up process. The error messages normally appear on the screen.
c) *Fatal errors*: These errors will not allow the computer to continue the boot-up process. If this occurs, usually the computer will communicate the errors through a series of audible beeps (*Table 19.0*). In this case, the computer needs to be repaired by a qualified technician.

POST ERROR BEEP TABLE	
Beeps	Error message
1	Refresh failure
2	Parity Error
3	Base 64K memory failure
4	Timer not operational
5	Processor error
6	Keyboard error
7	Processor exception interrupt error
8	Display memory read/write error
9	ROM checksum error
10	CMOS shutdown register read/write error
11	cache memory bad-do not enable cache

Table 19.0

- **Personal Computer Operating Systems and File Systems**

Different operating systems use different file systems. Some are designed specifically to work with more than one, for compatibility reasons; others work only with their own file system.

  o **MS-DOS:** *Microsoft's Disk Operating System (DOS)* uses *File Allocation File (FAT)* file system for file access. This included the *FAT12* and *FAT16* variations. *Microsoft's Windows 95* came with a "built-in" version of *DOS*, version 7.x. The *OEM Service Release 2 (OSR2)* version of Windows 95 supported *FAT32*, and therefore, *DOS* 7.1 that came with *Windows 95 OSR2* also supported *FAT32*. *FAT32* partition was not compatible with standard, pre-7.x DOS.

  o **Windows 3.x:** *Microsoft's Windows 3.x* was not a true, independent multi-tasking *Operating System*. It ran on top of *DOS* and used many *DOS* utilities and routines for disk access. Therefore, it used the same *FAT* file system that *DOS* did. The last version of *Windows 3.x*, *Windows for Workgroups 3.11*, included an enhancement called *"32-Bit File Access"*. This feature referred to the use of 32-bit protected mode routines for accessing the disk, instead of using the standard 16-bit *DOS* routines. In fact, this was the first implementation of the *VFAT* file system used by *Windows 95*, although not all of the *VFAT* features were included; only the use of 32-bit access routines. The difference was how to access the disk; the file system structures are "plain" *FAT*.

  o **Windows 95:** *Microsoft's Windows 95* has its own way of handling access to the hard disk, but in other ways it resembles, and even uses, standard *DOS*. *Windows 95* strives for performance while retaining compatibility with old *DOS* and *Windows 3.x* softwares. *Windows 95* in fact includes a version of *DOS*, *DOS 7.x*, that is designed to work with it and its file structures. The file system used is *VFAT*, which is supported by both *Windows*

225

95 and the *DOS 7.x.* Starting with *Windows 95 OEM Service Release 2 (OSR2)*, *FAT32* is also supported, which allows the use of larger hard disk partitions

o **Windows NT:** *Microsoft's Windows NT (NT for New Technology)* is another implementation of Windows that was designed from the ground up. Unlike the other Windows variations, *Windows NT* is not based on *DOS*. It supports two different file systems:

1. *NT file system (NTFS)*, the. This is an advanced UNIX-like file system.
2. *Regular FAT*, like that used by *DOS*. Support for *FAT* is included for compatibility reasons or for setting up PCs that *dual boot* to both *Windows NT* and another *Operating System*.

Even though *Windows NT* will read both *FAT* and *NTFS* partitions, they are not compatible with each other.

- **Dynamic Link Library (DLL)**

Dynamic Link Library (DLL) is best thought of as programmer's toolkits. They contain programming code that is often re-used from one application to another. Some *DLL*'s have one or two routines, while others may have a hundred or more. Rather than re-create the code, a programmer will use a *DLL* containing optimized code for the task. There are two types of *DLL*'s:

1. *Shared* or *common* – used by many programs.
2. *Proprietary* – used by one program.

The most common *DLL*'s are those that are part of a programming language such as C++, Visual Basic, etc. Those *DLL*'s provide the same "run time" environment as the developer's own system but do not allow for editing. The Windows Operating System typically allows one *DLL* to reside in active memory at any one time. The *DLL* will remain in memory until it is no longer needed.

- **Dynamic Data Exchange (DDE)**

Dynamic Data Exchange (DDE) is a Widows message protocol that lets applications request and exchange data.

- **Disk Structures and the Boot Process**

There are several major disk structures that are used to organize and control the storage of information on the PC using the *FAT (FAT32, VFAT)* file system. These structures are important to understand, since they control the way the hard disk works from a software perspective.

- **Active Partitions and Boot Managers**

Primary Partitions can be the only hard disk's partition set and used to boot the *Operating System*. Only one partition can be set bootable at a time because the *Master Boot Record (MBR)* does not know which volume's boot code has control of the boot process when the PC is turned ON. *DOS* calls the bootable partition the "active" partition. Partition on a new hard disk is created as a *Primary DOS Partition* using the standard *DOS* utility *FDISK*, then it is set active, if not then the *BIOS* will be unable to boot the *Operating System* (a typical error message "No boot device available" will appear on the monitor).

- **Partitioning, Partition Sizes and Drive Lettering**

Partitioning the hard disk is the act of dividing it into pieces; into logical volumes. This is one of the first things done when setting up a new hard disk, because partitions are one of the major disk structures that define how the disk is laid out. In fact, the hard disk must be partitioned, even if only into a single partition, before it is formatted and used. The choice of how the disk is partitioned is important because partition size has an important impact on both performance and on how efficiently the disk's space is utilized.

○ *FAT Sizes*

The *File Allocation Table* or *FAT* stores information about the clusters on the disk in a table. *FAT* file systems divide the hard disk into groups of bytes known as clusters then the file data into those clusters, bit by bit. Each cluster stores data from a single application or file only, so that when a program needs to retrieve a file, the *FAT* searches through all the clusters in order to find the one that holds the necessary data. There are three different varieties of this *FAT*, which vary based on their size. The system utility that is used to partition the disk will normally choose the correct type of *FAT* for the volume used, but sometimes a choice is given of which *FAT* to use. Since each *cluster* has one entry in the *FAT*, and these entries are used to hold the *cluster* number of the next *cluster* used by the file, the size of the *FAT* is the limiting factor on how many *clusters* any disk volume can contain. A *cluster* is the minimum amount of space that can be assigned to any file. No file can use part of a *cluster* under the *FAT* file system, i.e., the amount of space a file uses on the disk is "rounded up" to an integer multiple of the *cluster* size. If an application or a file does not contain enough data to use up all the bytes in a cluster, the remaining bytes in the cluster will not be used at all and the hard disk space left over in a cluster goes to waste. For example, if a cluster is 32 KB and a document file is 42 KB, the file will take up two full clusters (64 KB): 32 KB of the first and 10 KB of the second, the remaining 22 KB is wasted. The following are the three different *FAT* versions now in use:

- **FAT12:** The oldest type of *FAT* uses a 12-bit binary number to hold the cluster number. A volume formatted using *FAT12* can hold a maximum of 4,086 *clusters*, which is 2^{12} (minus 10 *clusters* to allow for reserved values to be used in the *FAT*). *FAT12* is therefore most suitable for smaller volumes, and is used on floppy disks and hard disk partitions smaller than about 16 MB.

- **FAT16:** The *FAT* used for most hard disk partitions uses a 16-bit binary number to hold cluster numbers. *FAT16* is referred to as *FAT* because it is the *de facto* standard for hard disks. A volume using *FAT16* can hold a maximum of 65,526 *clusters*, which is 2^{16} (less 10 *clusters* again for reserved values in the *FAT*). *FAT16* is used for hard disk volumes ranging in size from 16 MB to 2,048 MB (2 GB).

- **FAT32:** *FAT32* is supported by *Windows 95 OEM SR2* release, as well as *Windows 98*. *FAT32* uses a 28-bit binary cluster number (*not* 32) because 4 of the 32 bits are "reserved". 28 bits is still enough to permit large volumes. *FAT32* can theoretically handle volumes with over 268 million *clusters* ($2^{28} = 268,435,456$), and will support (theoretically) drives up to 2 TB (Terabytes) in size. However to do this the size of the *FAT* grows very large

Table 19.1 summarizes the three types of FAT compare:

Attribute	FAT12	FAT16	FAT32
Used For	Floppies and small hard disk volumes	Small to large hard disk volumes	Medium to very large hard disk volumes
Size of Each FAT Entry	12-bit	16-bit	28-bit
Maximum Number of Clusters	4,086	65,526	~268,435,456
Cluster Size Used	0.5 KB to 4 KB	2 KB to 32 KB	4 KB to 32 KB
Maximum Volume Size (bytes)	16,736,256	2,147,123,200	Approx. 2^{41}

Table 19.1

Table 19.2 shows cluster size of *FAT16* File System.

FAT16 File System	
Partition size (MB)	Cluster size (bytes)
0 – 32	512
33 – 64	1 K
65 – 127	2 K
128 – 255	4 K
256 – 511	8 K
512 - 1023	16 K
1024 - 2048	32 K

Table 19.2

Table 19.3 shows cluster size of *FAT32* File System.

FAT32 File System	
Partition size	Cluster size (bytes)
260 MB - 8 GB	4 K
8 GB - 16 GB	8 K
16 GB - 32 GB	16 K
greater than 32 GB	32 K

Table 19.3

Table 19.4 shows cluster size of *NTFS* File System.

NTFS File System	
Partition size (MB)	*Cluster size (bytes)*
512 or less	512
513 - 1,024	1 K
1,025 - 2,048	2 K
2,049 - 4,096	4 K
4,097 - 8,192	8 K
8,193 - 16,384	16 K
16,385 - 32,728	32 K
greater than 32,728	64 K

Table 19.4

o   *FAT Partition Efficiency: Slack*

One issue related to the FAT file system that has gained a lot more attention recently is the concept of *slack*. As larger and larger hard disks are being shipped on systems, especially low-end systems that are not being divided into multiple partitions, users have begun noticing that large quantities of the hard disk seem to "disappear". In many cases this can amount to hundreds of megabytes. The space does not really disappear. The space is simply wasted as a result of the cluster system that *FAT* uses. If a file containing exactly one byte, it will still use an entire cluster's worth of space. Then, that file can expand in size until it reaches the maximum size of a cluster, and it will take up no additional space. As soon as the file larger than what a single cluster can hold, a second cluster will be allocated, and the file's disk usage will double, even though the file may only increase in size by one byte. Since files always allocate whole clusters, this means that on average, the larger the cluster size of the volume, the more space that will be wasted. If a hard disk that has a truly random distribution of file sizes, then on average each file wastes half a cluster. The file sizes use any number of whole *clusters* and then a random amount of the last cluster, hence on average half a *cluster* is wasted. This means that if the cluster size of the disk is doubled, then the amount of storage that is wasted is also doubled. Storage space that is wasted in this manner, due to space left at the end of the last cluster allocated to the file, is commonly called *slack*. The situation is in reality usually worse than this theoretical average. The files on most hard disks do not follow a random size pattern, in fact most files tend to be small in size. A hard disk that uses more small files will in fact result in far more space being wasted. There are software utilities that can be used to analyze the amount of wasted space on the disk volumes. It is common for very large disks that are in single *FAT* partitions to waste up to 40% of their space due to *slack*.

For example, consider a 2.1 GB hard disk volume that is using 32 KB *clusters*. There are 10,000 files in the partition. Assume that each file has half a *cluster* of *slack*, this means that 16 KB (16,384 bytes) of space per file is wasted. Multiply that by 10,000 files, and get a total of 163 MB of *slack* space. Now assume that most of the files are smaller, therefore on average, each file has a *slack* space of about two-thirds of a *cluster* instead of one-half (21,845 bytes), this jumps to 218 MB of wasted space.

229

If unable to use a smaller *cluster* size for this disk, the amount of space wasted would reduce dramatically. *Table 19.5* shows a comparison of a 2.1 GB hard disk, the *slack* for various *cluster* sizes. The more files on the disk, the worse the *slack* gets. To estimate the percentage of disk space wasted,

bsize = size of hard disk in binary bytes

bsize =	2147483648			gsize =	2.15					
pcs (30%) =	0.33	*pcs (50%) =*	0.50	*pcs (60%) =*	0.67					
Cluster Size (KB)	*spf*	*Es full*	*Es in MB (pcs = 30%)*	*HDD % slack*	*Es in MB (pcs = 50%)*	*HDD % slack*	*Es in MB (pcs = 60%)*	*HDD % slack*
2	0.93	20	7	0.31	10	0.47	13	0.62
4	1.86	40	13	0.62	20	0.93	27	1.24
8	3.73	80	27	1.24	40	1.86	53	2.48
16	7.45	160	53	2.48	80	3.73	107	4.97
32	14.90	320	107	4.97	160	7.45	213	9.93

Table 19.5

pcs = percent cluster slack per file

$$gsize = size\ of\ hard\ disk\ in\ Gigabytes = \frac{bsize}{10^9}$$

$$spf = slack\ percentage\ factor = \frac{cluster\ size}{gsize}$$

$$Esf = Estimate\ slack\ Full = \frac{bsize * spf}{10^6 * 100}$$

$$Es = Estimate\ MB\ slack\ per\ hard\ disk = Esf * pcs$$

Example: A 2.1 GB hard disk,

$$bsize = 2 * 2^{30} = 2,147,483,648$$

$$gsize = \frac{2147483648}{10^9} = 2.15$$

$$pcs = \frac{2}{3} = 0.67$$

for 2 KB, the cluster size is 2, then spf = $\dfrac{2}{2.15} = 0.93$

$$Es\ Full = \dfrac{2147483648 * 0.93}{10^6 * 100} = 20$$

*Es (60%) = 20 * 0.67 = 13 MB*

The larger the cluster size used, the more of the disk's space is wasted due to slack. Therefore, it is better to use smaller cluster sizes whenever possible.

o ***Maximum Partition Size***

The number of clusters is limited by the nature of the *FAT* file system. There are performance tradeoffs in using smaller *cluster* sizes. Therefore, it is not always possible to use the absolute smallest *cluster* size in order to maximize free space. The place where partition size and cluster size interact is in the use of *FAT16*, the standard FAT type used for hard disks between 16 MB and 2,048 MB in size. The partitioning utility that is used to create the disk partitions will normally make the decision of what *cluster* size to use, by trying to use the smallest possible clusters in order to reduce slack. Since *FAT16* is limited in size to 65,525 *clusters*, each cluster size has a maximum partition size that it can support. If a partition is larger than that maximum, then the next larger cluster size is selected. *Table 19.6* shows the *FAT16* partition limits, the maximum cluster is 65,526 (*Table 19.1*). The maximum partition size is:

*bcs = binary cluster size = 2^{10} * cluster size*

mcs = maximum cluster size (FAT16 - Table 19.1) = 65526

*mpsd = maximum partition size (decimal bytes) = mcs * bcs*

mpsb = maximum partition size (binary bytes) = $\dfrac{mpsd}{2^{20}}$

For example: A cluster size 8K, *FAT16*

*bcs = 2^{10} * 8 = 8192*

*mpsd = 65526 * 8192 = 536788992*

mpsb = $\dfrac{536788992}{2^{20}} = 511.92\ MB$

Fat16 =65526			
Cluster Size (KB)	bcs	mpsd	mpsb
2	2048	134197248	127.98
4	4096	268394496	255.96
8	8192	536788992	511.92
16	16384	1073577984	1023.84
32	32768	2147155968	2047.69
64	65536	4294311936	4095.38

Table 19.6

Note:

- the maximum partition size for 8 KB clusters is 511.92, not 512 MB.
- Windows NT supports a 64 KB cluster size in FAT16, allowing a maximum partition of just under 4,096 MB. The amount of slack waste in a partition of this size is very large. The 64 KB cluster partition is not supported by Windows 95 or other FAT-using OS.

- ***Terminate and stay resident (TSR)***
A procedure that was used mostly during the DOS era of the computer operating system. It is a computer program that is copied to memory resident RAM, is run and remain active in memory.

- ***Configuration System file (Config.sys)***
The Config.sys file is a text file that was used primarily in DOS Operating System. This file would setup/specify the system parameters and the device drivers. As the computer system is starting up, the file executes first (after the BIOS is loaded) then the Autoexec.bat file.

- ***Automatic Executable Batch file (Autoexec.bat)***
The Autoexec.bat file is a batch file that was also used in DOS Operating System. This file as it is executed; it normally would list DOS commands to be performed in a sequence form.

- ***General Protection Fault Error (GPF Error)***
The GPF Error is a warning or an error from the MS Windows 3.1 or 3.11 that a program has tried to access a memory area but was not allowed to or was trying to perform a function but could not execute it.

- ***Bitmap***
A *bitmap* defines a display space and the color for each pixel or "bit" in the display space. A *GIF* and a *JPEG* are examples of graphic image file types that contain bit maps. A *bitmap* does not need to contain a bit of color-coded information for each pixel on every row. It only needs to contain information indicating a new color as the display scans along a row. Thus, an image with much solid color will tend to require a small bit map. Because a bit map uses a fixed or raster method of specifying an image, the image cannot be immediately rescaled by a user without losing definition. A vector graphic image, however, is designed to be quickly rescaled. Typically, an image is created using vector graphics and then, when the artist is satisfied with the image, it is converted to (or saved as) a raster graphic file or bit map. A *bitmap* is a file that indicates a color for each pixel along the horizontal axis or row (called the *x* coordinate) and a color for each

pixel along the vertical axis (called the *y* coordinate). A GIF file, for example, contains a *bitmap* of an image (along with other data). An image that has the specific record of each pixel (color, size and position) is called a *bitmap.*

- ***Joint Photographic Experts Group (JPEG)***

A *JPEG* is a graphic image created by choosing from a range of compression qualities (actually, from one of a suite of compression algorithms). When a JPEG is created or an image is converted from another format to a JPEG, the user is asked to specify the quality of image wanted. Since the highest quality results in the largest file, a trade-off between image quality and file size can be made. The *JPEG* scheme includes 29 distinct coding processes although a JPEG implementor may not use them all. Along with the *Graphic Interchange Format (GIF)* file, the *JPEG* is a file type supported by the World Wide Web protocol, usually with the file suffix of ".*jpg*". The *JPEG* is a color image graphics compression format named for the committee that designed the standard image compression algorithm. Using compression, computer files can be altered to take up less disk storage space. In the *JPEG* image compression system, some data may be sacrificed to achieve a high rate of data compression.

- ***Progressive JPEG***

A *progressive JPEG* is the *JPEG* equivalent of the *interlaced GIF.* It is an image created using the *JPEG* suite of compression algorithms that will "fade in" in successive waves of lines until the entire image has completely arrived. Like the interlaced *GIF*, a *progressive JPEG* is a more appealing way to deliver an image in these early days of relatively low bandwidth. As of mid-1996, not all browsers supported progressive *JPEGs.* However, *Netscape, Mosaic,* and *Microsoft's Internet Explorer* do support it.

- ***Tag Image File Format (TIFF)***

The *TIFF* format was developed in 1986 by an industry committee chaired by the *Aldus Corporation* (now part of *Adobe Software*). *Microsoft* and *Hewlett-Packard* were among the contributors to the format. *TIFF* is a common format for exchanging *raster* (*bitmapped*) images between application programs, including those used for scanning images. A *TIFF* file can be identified as a file with a ".*tif*" file name suffix. One of the most common graphic image formats, *TIFF* files are commonly used in desktop publishing, faxing, 3-D applications, and medical imaging applications. *TIFF* files can be in any of several classes, including gray scale, color palette, or *RGB* full color, and can include files with *JPEG, LZW,* or *CCITT* Group 4 standard run-length compression.

- ***Graphical User Interface (GUI)***

GUI originated at the *Xerox Palo Alto Research Laboratory* in the late 1970s. A *GUI* is a graphical (no text) user interface to a computer. The term came into existence because the first interactive user interfaces to computers were not graphical; they were text-and-keyboard oriented and usually consisted of commands that had to be remembered and computer responses to them. The command interface of the *DOS Operating System (OS)* is an example of the typical user-computer interface before *GUI.* An intermediate step in user interfaces between the command line interface and the *GUI* was the non-graphical *menu-based interface*, which let the user interact by using a mouse rather than by having to type in keyboard commands. Today, most *OS* provide a graphical user interface. Applications typically use the elements of the *GUI* that come with the *OS* and add their own graphical user interface elements and ideas. A *GUI* sometimes uses one or more metaphors for objects familiar in real life, such as the desktop, the view through a window, or the physical layout in a building. Elements of a *GUI* include such things as: windows, pull-down menus, buttons, scroll bars, iconic images, wizards, the mouse, etc. With the increasing use of multimedia as part of the *GUI*, sound, voice, motion video, and virtual reality interfaces seem likely to become part of the *GUI* for many applications. A system's *GUI* along with its input devices is sometimes referred to as its "*look-and-feel*". The *GUI* familiar to most of people today in either the Mac or the Windows OS and their applications. Apple used it in their first Macintosh computers. Later, *Microsoft* used many of the same ideas in their first version of the Windows Operating System for *IBM*-compatible PCs. When creating an application, many object-oriented tools exist that facilitate writing a *GUI*. Each *GUI* element is defined as a class from which a user can create object instances for the application. The user can code or modify prepackaged methods that an object will use to respond to user stimuli.

- ***Graphics Interchange Format (GIF)***

A *GIF* is one of the two most common file formats for graphic images on the *World Wide Web*. The other is the *JPEG*. On the Web and elsewhere on the Internet (for example, *Bulletin Board Services or BBS*), the *GIF* has become a de facto standard form of image. The format is owned by *Compuserve*. Technically, a *GIF* uses the 2D *Raster* data type, is encoded in binary, and uses *LZW* compression. There are two versions of the format, 87a and 89a. Version 89a (July, 1989) allows for the possibility of an animated *GIF*, which is a short sequence of images within a single *GIF* file. A *GIF89a* can also be specified for interlaced presentation.

- ***Lempel-Zif-Welsh (LZW)***

LZW is a popular data compression technique developed in 1977 by *J. Ziv* and *A. Lempel*, and later refined by *T. Welsh*. The patent for *LZW* is owned by *Unisys Corporation*. It is the compression algorithm used in the *GIF* graphics file format, which is one of the standard graphic formats used by the *World Wide Web*.

- *GIF89a*

A *GIF89a* graphics file is an image formatted according to *GIF* Version 89a (July, 1989). There was an earlier Version, 87a, from May of 1987, but most images that are seen on the *World Wide Web* have probably been created in the newer format. One of the chief advantages of the newer format is the ability to create an animated image that can be played after transmitting to a viewer page.

- ***Animated GIF***

An *animated GIF* is a graphic image on a Web page that moves - for example, a twirling icon or a banner with a hand that waves or letters that magically get larger. In particular, an *animated GIF* is a file in the *Graphics Interchange Format* specified as *GIF89a* that contains within the single file a set of images that are presented in a specified order. An *animated GIF* can loop endlessly (and it appears as though the document never finishes arriving) or it can present one or a few sequences and then stop the animation.

- ***Portable Network Graphics (PNG)***

PNG is a file format for compressed graphic images that, in time, is expected to replace the *GIF* format that is widely used on today's Internet. The *GIF* format, patented by *Compuserve* (now owned by America Online), and its usage in image-handling software involves licensing or other legal considerations. (Web users can make, view, and send *GIF* files freely but they cannot develop software that builds them without an arrangement with Compuserve.) The *PNG* format, on the other hand, was developed by an Internet committee expressly to be patent-free. It provides a number of improvements over the *GIF* format. Like a *GIF*, a *PNG* file is compressed in loss-less fashion (meaning all image information is restored when the file is decompressed during viewing). A *PNG* file is not intended to replace the *JPEG* format, which is "lossy" but lets the creator make a trade-off between file size and image quality when the image is compressed. Typically, an image in a *PNG* file can be 10 to 30% more compressed than in a *GIF* format.

The *PNG* format includes these features:

- One color transparent can not only be made, but the degree of transparency can be controlled (this is also called "*opacity*").

- Interlacing of the image is supported and is faster in developing than in the GIF format.

- *Gamma correction* allows the user to "tune" the image in terms of color brightness required by specific display manufacturers.

- Images can be saved using true color as well as in the palette and gray-scale formats provided by the GIF.

Unlike the *GIF89a*, the *PNG* format does not support animation since it can not contain multiple images. The *PNG* is described as "extensible", however. Software houses will be able to develop variations of *PNG* that can contain multiple, scriptable images.

- **interlaced GIF**

An *interlaced GIF* is a *GIF* image that seems to arrive on the PC display like an image coming through a slowly-opening Venetian blind. A fuzzy outline of an image is gradually replaced by seven successive waves of bit streams that fill in the missing lines until the image arrives at its full resolution. Among the advantages for the viewer using high speed modems (28.8 Kbps or faster) are that the wait time for an image seems less and the viewer can sometimes get enough information about the image to decide to click on it or move elsewhere.

- **Raster Graphics**

Raster graphics are digital images created or captured (for example, by scanning in a photo) as a set of samples of a given space. A *raster* is a grid of *x*- (horizontal) and *y*- (vertical) coordinates on a display space. (for three-dimensional images, a *z*-coordinate). A *raster image* file identifies which of these coordinates to illuminate in monochrome or color values. The raster file is sometimes referred to as a bitmap because it contains information that is directly mapped to the display grid. A *raster file* is usually larger than a vector image file and difficult to modify without loss of information, although there are software tools that can convert a raster file into a vector file for refinement and changes. Examples of raster image file types are: *BMP*, *TIFF*, *GIF*, and *JPEG* files.

- **Vector Graphics**

Vector graphics is the creation of digital images through a sequence of commands or mathematical statements that place lines and shapes in a given two-dimensional or three-dimensional space. In physics, a *vector* is a representation of both a quantity and a direction at the same time. In vector graphics, the file that results from a graphic artist's work is created and saved as a sequence of vector statements. For example, instead of containing a bit in the file for each bit of a line drawing, a vector graphic file describes a series of points to be connected. One result is a much smaller file. At some point, a vector image is converted into a raster image, which maps bits directly to a display space (and is sometimes called a *bitmap*). The vector image can be converted to a raster image file prior to its display so that it can be ported between systems. A vector file is sometimes called a *geometric* file. Most images created with tools such

Carlos E. Hattab

as *Adobe Illustrator* and *CorelDraw* are in the form of vector image files. Vector image files are easier to modify than raster image files (which can, however, sometimes be reconverted to vector files for further refinement). Animation images are also usually created as vector files. Shockwave's Flash product lets the user create 2-D and 3-D animations that are sent to a requestor as a vector file and then rasterized "on the fly" as they arrive.

- *Software bug*

A software bug is a mistake in the design or implementation of a computer's program. That mistake causes incorrect results.

- Computer crash

A computer crash is an unexpected interruption or shutdown of a working computer.

- *Software Program*

A software program is a set of instructions that command the computer to perform a task. It is not only a set of executable piece of computer language code, but also is simultaneously a complete set of technical documentation for the software system that it represents. The text of a program should meet very high standards for readability as well as performance. By the time the solution to an algorithmic problem has been fully developed and tested, the computer software language code, for example *PASCAL*, should be reasonably self-documenting when it is read by someone unfamiliar with the problem but familiar with the methodology and the application domain from which the problem is taken. That is, the text of a program -- with its documentation, stepwise structure and pre- & post-conditions – should be easy to read for a professional colleague as an article in a newspaper would be for a reader. There are three levels of language:

a) The *PASCAL* code.
b) The process description, which states in a language (English) what is happening at each step of the process that the program describes.
c) The assertions that describe the state of the computation before and after various steps in the process are executed.

Together, these three levels contain a lot of redundancy; but they express the same goals or ideas in three different styles:

a) Formal algorithmic (*PASCAL* software language).
b) Informal algorithmic (plain English description).
c) Formal declarative (specifications),

This redundancy emphasizes the fact that a programmed solution to an algorithmic problem is written to serve the interests of three different audiences:

a) The computer, which executes the program.
b) The authors, who design and implement the program.
c) The readers, who maintain and verify the program's correctness.

The computer is interested in the *PASCAL* language code; it will follow the code precisely as it carries out the steps of the process. The authors are interested in the Pascal code, the annotated English commentary, and the pre- & post-conditions, so that they can correctly develop, understand, and later modify or improve the program. The readers are interested in the *PASCAL* code and the accompanying pre-conditions, post-conditions and even sometimes the embedded assertions, so that they can develop a rigorous demonstration (verification) of the program's correctness.

A sample software program in *PASCAL* language for the three audiences:
Program *CalculateGPA;*

{This PASCAL program calculates the Grade Point Average (GPA) for a series of one or more numeric grades, each in the range of 0 to 4, entered at the keyboard of a computer as a list.}

uses

Lists;

var

{Define the variables used in this program.}

Grades : Lists;

i, n : integer;

Sum, GPA : real;

Begin

{PRE- : input = (grades [1], grades[2],...grades[n]), n > 0 and for all i in [1,2,...,n]: grades[i] in [0...4]}

{Step 1. Obtain the input list of grades}

Writeln ('Enter a list of grades:');

Readlist (Grades);

{Step 2. Compute n = the number of grades in the list.}

n:=lenghtlist(Grades);

{Step 3. Compute the sum = the sum of the grades in the list.}

Sum :=0;

For i := 1 to n do

Sum := Sum + Grade(i);

{Step 4. Compute GPA = Sum / n.}

GPA := Sum / n;

{Step 5. Display GPA.}

Writeln('The GPA of these grades = ',GPA:5:2);

{Post-: Input = empty and

Output = (Sum i in [1,..,n]: grades [i] / n)

END. *{CalculateGPA}*

CHAPTER XX
SELECTING A COMPUTER

The selection of a computer system will involve many aspects. Besides the *MIPS* and the *MFLOPS*, there are other considerations for PC performance, characteristics and power measurement tools that need to be addressed, such as:

1. ***CPU Throughput:*** It is a measure of the number of tasks and/or requests that a CPU can execute per unit of time. Example: the throughput of *Pentium* II-450MHz is about 50 Mbps (for some applications).

2. ***CPU Utilization:*** it is the fraction of time that a CPU is busy executing programs. It is the ratio of busy time and total elapsed time over a given period.

3. ***Response Time or Turnaround Time:*** It is the time interval between the time a request is issued for service and the time that the service is completed.

4. ***Memory Bandwidth:*** It is the number of memory words that can be accessed per unit time.

5. ***Memory Access Time:*** It is the average time that it takes the CPU to access the memory. Example: SIMM RAM access time is 60 nanoseconds.

6. ***Memory Size:*** It is the capacity of the memory, which is an indication of the volume of data that the memory can hold. Example: a PC with 32 Megabytes of RAM.

7. ***Generality:*** It is a measure that determines the range of applications for a system's architecture. Example: Scientific architecture would prefer using scientific applications instead of industrial one. Example: this system can open a document, perform a math calculation, attach to a network, etc…

8. ***Ease of Use:*** It is a measure of how easy a computer (including Operating Systems and application software) it is to use. Example: after power is turned ON, what to do.

9. ***Expandability:*** It is a measure of how easy it is to expand/add to the capabilities of the computer system's architecture. Example: RAM memory, hard disk, etc… Example: flexibility to increase the RAM size or add another hard disk drive.

10. ***Compatibility:*** It is a measure of how compatible the architecture is with the existing or previous computer system. Example: as the new version of application has been installed, the old document (for earlier version) can still be accessed.

11. ***Reliability:*** It is a measure of the probability of faults (crashes) or the mean time between failures or errors. Example: the system can run one or more application without shutting down.

12. ***Ergonomics:*** It is the study of human performance and its application to the design of technological systems. The goal of this activity is to enhance productivity, safety, convenience and quality of life. Example: the human-computer interface issues (monitor location, keyboard feel, mouse reach, etc…).

CHAPTER XXI
THE COMPUTER VIRUS

A virus is a unique computer program specifically developed to attach or associate itself with another existing computer program in a matter that whenever the existing program is run and active, the virus program is active as well. It is a program designed to destroy data or halt operations on the computer system by copying itself onto the files and executing when those files are loaded active into memory. Sometimes, a virus program replicates itself as it is attached to other existing computer programs, other times; it replaces the existing computer programs. In all cases, the existing computer program is now "infected", and this infection can be deadly because the virus can infect any computer program ranging from the "disk boot" area to the application software and finally to the personal documents.

Figure 21.0

Computer viruses are small programs designed to attach themselves to a computer, running discretely and spreading themselves to "infect" other systems. Sometimes malicious, other times just annoying, they are always a concern to the modern computer user. Virus computer programs are not accidently installed on a PC; computer programmers intentionally develop them. In most cases, virus computer programs are not as harmful or deadly as they are known to be but some of them can be very destructive. They can cause damage to some files by replacing or destroying part or all of the files' data. Computer virus program can perform some harm or destruction to a computer performance by occupying a space in the storage media disk (fixed: hard disk or removable: floppy disk) or memory RAM, by using the CPU resources such as interrupts. Virus programs can be annoying and/or invisible to most computer users. Some of the many types of viruses are:

1. ***Boot sector virus*** – this virus tends to infect and damage the specific area of the hard disk: Boot files. Another subtype of this is the *Master Boot Record (MBR)* virus, which infects the *MBR*. This is normally deadly to the computer system, in most cases; a

complete reformat of the hard disk is required to recover from the infection. The boot code that is stored on the hard disk and executed when the disk is booted up, is a prime target for viruses. The reason is simple: the goal of the virus writer is to get the virus code executed as often as possible, to allow it to spread and cause whatever other mischief it is written to create. One of the two major classes of viruses, called *boot sector infectors*, targets the vulnerable boot areas of the hard disk. The other major group of viruses attacks individual files. Some infect the code in the *MBR* while others infect the code in the volume boot sector(s). By infecting this code, the virus assures itself of always being able to load into memory when the PC is booted, as long as the PC boot from the volume that is infected. There is always code in any disk (hard or floppy) that is formatted, whether or not the system files are present on the disk.

2. ***File infectors virus*** – also known as parasitic virus. This virus penetrates the computer system by attaching itself onto an executable infected program (exe, com). It then copies itself onto the computer system memory (*RAM*). From there, it replicates itself to another computer program that becomes active in memory. This virus can damage part or the entire file.

3. ***Macro virus*** – Macro is a computer script program that assists users with a specific application. The *Macro* virus is also similar type script program that attacks the macro program. If a document has a macro program, this virus can take advantage and attaches itself onto this document, once there; it can do all the damages freely. If there are other documents in the computer system with Macro, this virus can replicate itself and do the damages to those documents as well.

In most cases, viruses can penetrate a computer system through an infected program file (com, exe, boot sector) not a document or a graphics or a sound file. Before the Internet era, computers were infected with viruses through a removable media or *Bulletin Board* file download. Today in most cases, viruses come from removable media, Intranet, Internet or a network. *Electronic mail* (*E-mail*) messages can contain a virus that can attach an infected computer program. A virus is a computer program that must be run to become active. In an e-mail, unless the attached infected program is downloaded and executed, the virus is not active. A typical way for the virus to transfer from an infected program to another non-infected program is through the computer memory. When an infected computer application program is run, the virus copies itself onto the system memory (*RAM*). If the virus executes its code and infects another computer program immediately, it is called a *direct-action* virus. Even when application is terminated, the virus computer program remains in the system memory active; it is now called *resident* virus. From there, this virus program can copy itself onto other programs in the computer, whether the computer CPU is idle or not.

As a result of this, anti-virus computer programs were developed to detect and destroy them. This anti-virus program now has to be active at all times, hence, it consumes *RAM* memory and CPU resources. Either way, the computer's performance is slowed down. With an anti-virus computer program, a user can avoid infection by a virus by disinfecting floppy disks and files that are introduced to the computer.

Figure 21.1

- **Malware**

In general, virus computer programs are part of computer category software known as *malware or malicious-logic-software.* Virus computer programs, in most cases, are invisible and they are copied from one computer to another without the user's knowledge. They are intentionally developed to execute and perform some specific function within the computer's system.

There are other kinds of *malware:*

1. The *Worm* is a computer program that replicates itself but does not "infect" other computer programs. The Worm can copy itself multiple times within a computer system (media storage: floppy disk, hard disk, etc.) or within a network of computers.
2. The *Trojan Horse* is a computer program that is attached and hidden in an application program. Unlike the *Worm*, the *Trojan Horse* does not copy itself but as the application program is run (becomes active), this virus is activated and may perform some unwanted actions that could disrupt the application and eventually shut down the computer.
3. The *Dropper* is a computer program that is designed to infect an application program while avoiding detection by "anti-virus" computer program.

Malware computer programs can be very destructive inside a computer. They can delete a part of an application, personal documents, templates, etc...

The best method to protect a computer system from virus attack is to learn and understand about the virus program. A new way of copying and download a file must be used and followed at all time. An anti-virus program is absolutely needed but it has to be continuously updated so that its database can have the newest virus program pattern. Virus programs are generated daily, only the deadly ones are announced or talked about. Any user must be careful when copying or executing a computer program. Most of the anti-virus programs can check a computer program file for viruses prior to executing that file. In some cases, if a file is suspiciously infected with a virus, then the anti-virus programs alert the user about the virus. Among the choices is to clean or delete the infected file from the media storage. If a backup is available, it is best to delete the

241

infected file. It is recommended to regularly run anti-virus program, scan and defragment the media storage.

In conclusion, as long as there is a computer program, there will be a virus program.

CHAPTER XXII

CONCLUSION

In essence, the computer system does all the simple and complex functions and computations that the human brain can do, but its great power lies in its speed and accuracy. The computer will not carry out any process, which the human brain is unable to perform, but it will solve tasks or problems, which involve handling simultaneously hundreds of the processes. The brain solves them one at a time. Furthermore, the computer system forgets nothing; it is programmed to remember tasks and results, to retrieve large data files in seconds, to transmit a message to a satellite, to perform a backup of specific information at 2:00 am.

Computers are designed and programmed to assist the human with tasks. In reviewing the architecture of the computer system, as shown in *Figure 22.0*, the difference between mainframe computers, minicomputers, microcomputer and the Personal Computers does not lie in the fundamental block used to build the computer; instead, it relates to the performance and capacity of the electronics circuitry used to implement these blocks in the design and development.

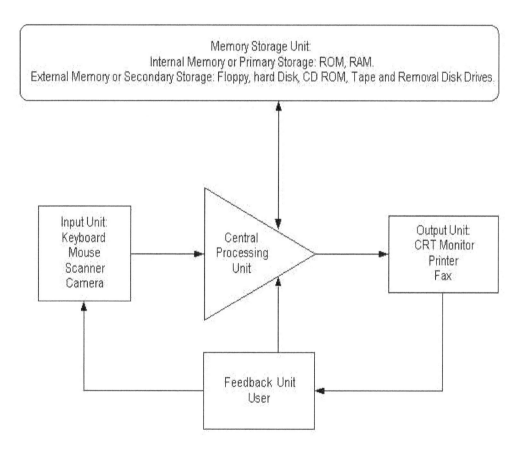

Figure 22.0

243

Carlos E. Hattab

APPENDIX A
PROCESSORS REFERENCE

The following tables are some of the CPUs introduced (other than Intel) by:

1. Advanced Micro Designs (AMD).

CPU	Clock Speed	Socket
Am486DX4-100	100	Socket 1,2,3
Am486DX4-120	120	Socket 1,2,3
Am5x86	75	Socket 1,2,3
K5 PR75	75	Socket 5,7
K5 PR90	90	Socket 5,7
K5 PR100	100	Socket 5,7
K5 PR120	90	Socket 5,7
K5 PR133	100	Socket 5,7
K5 PR166	116.66	Socket 5,7
K6-166 MMX	166	Socket 7
K6-200 MMX	200	Socket 7
K6-233 MMX	233	Socket 7

Table A.0

2. Cyrix.

CPU Type/Speed	Clock Speed	CPU Clock	Motherboard Speed
6x86-PR120	100	2x	50
6x86-PR133	110	2x	55
6x86-PR150	120	2x	60
6x86-PR166	133	2x	66
6x86-PR200	150	2x	75
6x86MX-PR166	150	2.5x	60
6x86MX-PR200	166	2.5x	66
6x86MX-PR233	188	2.5x	75

Table A.1

3. IBM computer systems.

Date Introduced	Computer Series/Model	Features / Comments
1953	700	Vacuum tubes.
1954	650	Magnetic-drum storage.
1960	1400	Oriented to business use.
	1620	Oriented to scientific use.
	7000	Transistors, Oriented to business and scientific use.
1962	1130	Integrated Circuits, small, special-purpose.
1963	1800	Integrated Circuits, small, special-purpose.
1965	360	Oriented to business and scientific use.
1969	System/3	Midi-Small computer.
1970	System/7	Replacement for 1800 series.
1973	370	IBM's most popular system.
	System/38	Powerful, general purpose, supporting extensive data bases.
1975	System/32	Small system for business use.
	5100	Portable computer.
1976	Series/1	Small system for business use.
1977	370-3031	Extends capabilities of 360 series.
	System/34	Versatile small system for experienced users.
1978	5110	Small system for business use.
1980	370-3033	Extends capabilities of 360 series.
	5120	Small system for business use.
1981	Datamaster	Small system with data and word processing.
	Personal Computer	Microcomputer for home and office use.
1983	S/36	Small system for business use.
1985	3090	IBM's most powerful processors system.
1986	9370	Mid-range departmental system.
1987	PS/2	New generation of Microcomputers.
1988	AS400	Replaced the S/36 and S/38; relational data base system; Middle range family of computers based upon a relational data base system.

Table A.2

APPENDIX B
CPU SOCKET REFERENCE

The followings are the pin-out infrastructure of Sockets 1 through 6.

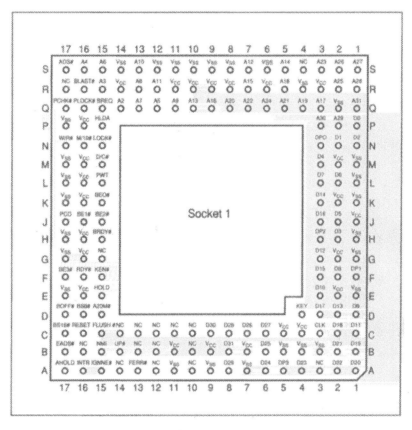

Socket 1 pin-out
Figure B.0

Carlos E. Hattab

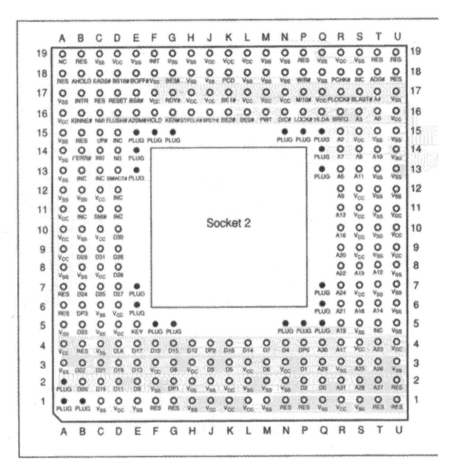

Socket 2 pin-out
Figure B.1

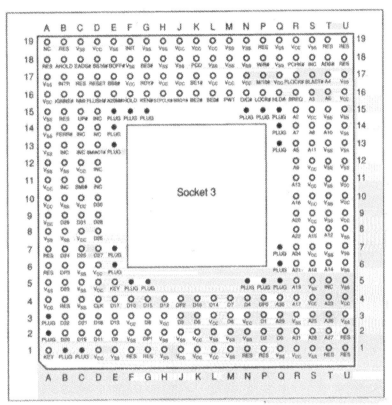

Socket 3 pin-out
Figure B.2

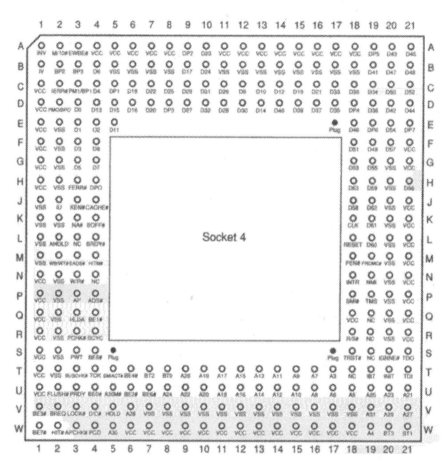

Socket 4 pin-out
Figure B.3

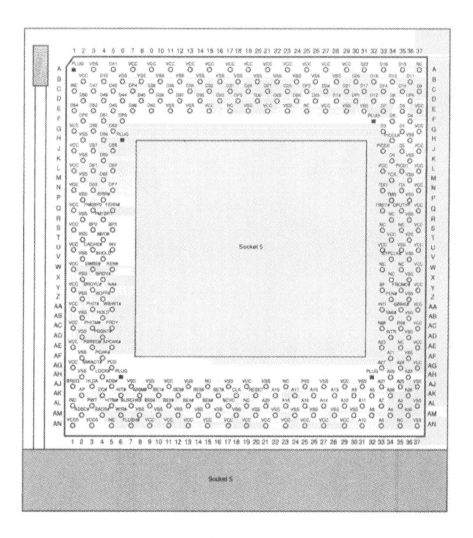

Socket 5 pin-out
Figure B.4

251

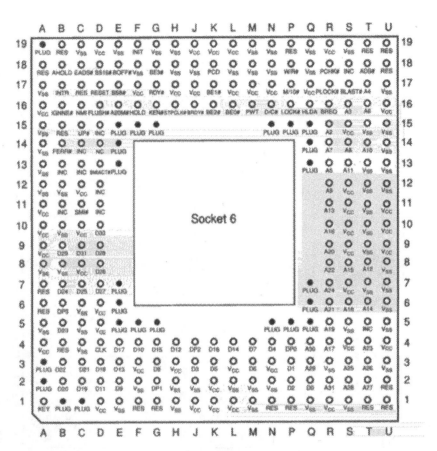

Socket 6 pin-out
Figure B.5

GLOSSARY

A Programming Language (APL): A general purpose computer language using special-purpose characters. This is a terminal-oriented high-level programming language that is especially suitable for interactive problem solving.

Abacus: A simple ancient device used for mathematical calculations. It consists of a rectangular frame with beads strung on wires.

Access: To get or retrieve data or information from a computer system.

Access limits: A method used to restrict computer users in accessing protected data or information, for example, read access, full access, etc.

Access mechanism: The device that positions the Read/Write Head of a direct-access storage device over a particular track.

Access, random: The ability to get information from a storage device in any sequence.

Access, serial: The ability to get information from a storage device only in the sequence in which it was stored.

Access speed: The rate at which information can be moved from or to a computer's memory.

Access time: The average amount of time required to retrieve a quantity of data from its storage location. It is the amount of time it takes from the start of a random read operation until the data starts to be read from the disk.

Accelerated Graphics Port (AGP): *AGP* is based on *PCI* Bus, but is designed especially for the throughput demands of 3-D graphics.

Accurate data: An element of data integrity guaranteeing the user's confidence in the results of data processing by assuring that the source of the data to be entered is reliable.

Acoustic coupler: A device that converts electrical signals into a sound, to allow a computer to communicate via telephone line and vice-versa.

Acoustic coupler modem: A device used in telecommunications that is attached to a computer by a cable and connects to a telephone handset.

Active backplane: system has some circuitry mounted on the backplane but not a CPU.

Ada: A high-level programming language named for Augusta Ada Byron, Countess of Lovelace. Ada is a sophisticated structured language that supports concurrent processing.

Adapter: It is a physical device that allows one hardware or electronic interface to be adapted (accommodated without loss of function) to another hardware or electronic interface.

Address: A unique identifier assigned to each memory location within the primary storage.

Address Bus: It transfers the information to a specified location, i.e. to where the data should go.

ALGOL: An international computer language widely used to publish computer algorithms.

Algorithm: A well-defined mathematical or logical procedure for solving a problem in a finite number of processing operations.

Alphanumeric characters: Any letters, numbers or special symbols (such as punctuation marks) used in data processing operations.

I

American Standard Code for Information Interchange (ASCII): A seven-bit standard code used for information interchange among data processing systems, communications and associated equipment.

ASCII-8: An eight-bit version of ASCII developed for computers that require eight-bit rather than seven-bit code.

American National Standards Institute (ANSI): The authority for establishing industrial standards in the U.S.A. The *ANSI* is the primary organization for fostering the development of technology standards in the United States.

Ampère (André Marie Ampère, 1775-1836): The current (I) that would produce a force of 2 * 10^{-7} Newton per meter of length between two parallel conductors one meter apart. It is equal to a charge flow of 1 coulomb per second. The symbol is *A*.

AND gate: The *AND* gate logic output is equal to the *AND* product of the logic inputs.

Analog computer: A computer that operates on data in the form of continuously variable physical quantities rather than digits. It is a computer that measures the change in continuous electrical or physical conditions rather than counting data; contrast with digital computer.

Analog transmission: Transmission of data over communication channels in a continuous wave form.

Analog quantity: It is a quantity that may assume any numerical value within the range of possible output or outputs, it is a continuously variable quantity; one that may assume any value within a limited range.

Analytical Engine: A machine, designed by Charles Babbage, capable of addition, substraction, multiplication, division and storage of intermediate results in a memory unit.

Application software: A software which performs an application-related function on a computer.

Arithmetic Logic Unit (ALU): It performs arithmetic and logical operations.

Artificial Intelligence (AI): The capability of any device to perform functions that are normally associated with human intelligence, such as reasoning, learning and self improvement.

Asynchronous Computer: A digital computer that takes a variable amount of time to complete each processing operation, depending on its intrinsic complexity.

Artificial Intelligence (AI): The ability of a computer to perform tasks that most people know that it would require intelligence.

Arithmetic Logical Unit (ALU): The section or part of a computer that performs arithmetic and logical operations.

Assembler: A computer program that converts symbolic instructions written by people into binary (machine language) to run on a computer.

Asynchronous transfer (ASYNC): A method of transferring data that requires that the Bus wait for a *Request-Acknowledge* handshake for each byte of data.

AT Attachment (ATA): The *ATA* is a disk drive implementation that integrates the controller on the disk drive itself.

AT Attachment Packet Interface (ATAPI): The *ATAPI* is an extension to *EIDE* (also called *ATA-2*) that enables the interface to support *CD-ROM* players and tape drives.

Atanasoff-Berry-Computer (ABC): The first electronic digital computer that was designed and built by Dr. John Atanasoff and Clifford Berry.

Automatic: Any device that, under certain conditions, can function without intervention by a human operator.

Automation: The replacement of human labor by a machine. The implementation of a manufacturing process by automatic means using mechanical, electromechanical, or electronic devices.

Auxiliary storage devices: Devices used to store data and instructions when they are not being used in the main computer memory. Example: magnetic disk or magnetic tape.

»»»»»»- B -«««««««

Background program: A low-priority program that is executed whenever the computer's resources are not required to process higher-priority programs.

Backup: It is the safety net of a total PC data loss.

Backup file: A copy of a file, typically stored on magnetic tape.

Backplane: systems use a main card with slots to plug all of the PC components into.

Backwards-compatible: An older device can be attached to a newer computer with support for a newer standard, the older device will work at the older and slower data rate.

Basic Input/Output System (BIOS): It is a special program of software stored in most computer memory that controls the startup process of the computer.

Batch file: A file which can contain statements to cause multiple programs to execute without an operator intervention.

Batch processing: A mode for processing data on a computer whereby data records are accumulated and are processed as a group when all of the required records have been gathered.

Beginner's All-Purpose Symbolic Instruction Code (BASIC): A programming language developed by *Dr. John Kemeny*, initially used by *Dartmouth College* students, then later by millions. BASIC is a simple procedure-oriented computer language widely used in educational institutions, small business applications and in hobby computing.

Benchmark: It is a test used to compare performance of the hardware and/or software.

Binary: It is the simplest number system that uses positional notation. The binary is base 2, it contains only two elements or states; therefore the digits can only be a zero (0) or a one (1).

Binary digit: The basic unit for storing data in main computer memory.

Binary Coded Decimal (BCD): Can also be referred to as a nibble, it is a group of four bits associated together.

Bistable device: A device with exactly two stable states, such as ON or OFF.

Bit: It is a single binary digit such as 0 or 1. The term is derived from the contraction of *b*inary dig*it*.

Bit-mapped displays: Display monitors, also called dot-addressable displays, used for graphics.

Bitrate denotes the average number of bits that one second of audio data will consume. **Bits-per-inch (bpi)**: A measurement of the recording density of a diskette.

Black box: An electronic or mechanical device that alters its inputs signals in a predictable way but whose inner workings are often a mystery to the user.

Blinking: A feature of a screen that allows characters or words to blink.

Boolean algebra: It is a quantity that may, at different times, be equal to either logic 0 or logic 1.

Boot-up: The process of loading an Operating System into main computer memory.

Buffer: A high-speed, low capacity storage register used to increase the efficiency of data transfer operations between two storage units of grossly different data transfer rates.

III

Bug: A mistake or malfunction in a program.

Burst EDO (BEDO) DRAM: It is a new type of *EDO DRAM* that can process four memory addresses in one *burst*.

Burst Rate: The data are buffered and sent in a burst at maximum speed over the bus, so the bus is utilized for only a short period of time. It is then released and can be used by other devices.

Bus: It is a collection of parallel wires (connections or traces on *PCB*) through which electrical (*voltage*) or electronic signals (*data*) is transmitted from one part of a computer system to another.

Bus mastering: A peripheral controller which is capable of transferring data over the host bus without the host CPU's assistance, is called a bus master. Bus mastering delivers maximum relief of the host on I/Os, but it requires extra intelligence on the peripheral controller.

Byte: A unit of information that holds one letter or digit.

»»»»»»- **C** -«««««««

C programming language: A general purpose, portable programming language featuring concise expression of functions to be performed, and a design that permits writing well structured programs.

Cache: The cache memory, also known as *high-speed buffer*, is a high-speed memory bank set aside (reserved by the computer) for frequently accessed data or information.

CAD/CAM: An abbreviation for Computer-Aided-Design and Computer-Aided-Manufacturing.

Calculator: A mechanical or electronic data processing device that requires frequent intervention by a human operator.

Capacitor: It is used to store electrical charge for a period of time then releases it. Capacitor unit is Farad (F).

Card punch: A mechanical device that punches holes into a set of cardboard cards to represent numbers, letters, and special characters.

Card reader: A device that senses the locations of the holes in a deck of punched cards and translates them into a machine-processable code.

Card sorting: The process of separating a stack of punched cards into separate classifications in accordance with the holes punched into the individual cards.

Cathode Ray Tube (CRT): An output unit with a TV like screen used to display information from a computer. The CRT is a vacuum tube, similar to a television picture tube, in which an electron gun, controlled by electromagnets, directs a narrow beam of electrons onto specific regions of a fluorescent screen or photographic surface.

Central Processing Unit (CPU): The electronic circuits which, as a part of the processor unit, cause processing to occur by interpreting and executing instructions and controlling the input, output, and storage operations. It is the brain of the computer system.

Character: Any symbol, digit, letter, or punctuation mark stored or processed by a digital computer.

Character reader: An input device that reads alphanumeric characters directly from a printed document.

Charged-Coupled Device (CCD): A volatile, semi-random-access storage device in which the binary digits held in storage are continuously circulated around a number of continuous electronic loops, each of which is composed of a chain of Solid-State switches.

Chip: It is a small piece of a semiconductor material (typically silicon) on which an IC is embedded.

Circuit: A complete electrical pathway designed for the controlled flow of electrons.

Clock cycles: The CPU requires a fixed number of clock ticks (or *clock cycles*) to execute each instruction.

Clock-pulse circuit: A circuit that generates a series of evenly spaced timing pulses to help orchestrate the operations carried out by a synchronous digital computer.

Clock rate: Also referred to as **Clock speed**.

Clock speed: Also referred to as **clock rate**, is the speed at which a CPU of a computer executes program instructions.

Command: It is an instruction that an initiator issues to a target specifying the task to be carried out.

Common Business Oriented Language (COBOL): A widely used machine-independent business programming language that allows programs to be written in an English-like fashion.

Communication link: A physical method of connecting one location to another for the purpose of transmitting and receiving information.

Compact Disc, Read Only Memory (CD-ROM): Introduced in the 80's as a media device of the PC, the *Compact Disk (CD)* is an optical disk.

Compiler: A program, often called a translator, that interprets computer programs written in a high-level language and converts them to machine language.

Complementary Metal Oxide Semiconductor Logic (CMOS): The *CMOS* logic is popular because of low power dissipation, hence low heat.

Computer: An electronic device, operating under the control of instructions stored in its own memory unit, which can accept and store data, perform arithmetic and logical operations on that data without human intervention, and produce output from the processing.

Computer-Aided Design (CAD): A process involving direct real-time interaction between a designer and a computer, usually by means of a CRT and a light-pen or digitizer pad.

Computer-Aided Manufacturing (CAM): A manufacturing system that uses digital computers extensively to operate robots and other manufacturing machines for maximum practical output.

Computer analyst: A person who studies potential computer applications and describes how they will work.

Computer graphics: Displaying information in the form of charts, graphs, or pictures. Information can easily be understood.

Computer hardware: The input, processor, output and auxiliary storage units which process data.

Computer program: A series of instructions which directs a computer to perform input, arithmetic, logical, output and storage operations.

Computer programmers: People who design, write, test and implement programs on a computer.

Computer science: A field of study to become systems programmers, software specialists or research computer scientists.

Computer software: A series of instructions, also called programs, which directs a computer to perform input, arithmetic, logical, output and storage operations.

Computer technology: A field of study that deals with computer hardware and the electronic components comprising computer hardware.

Constant Angular Velocity (CAV): In a CD-ROM drive, *CAV* is regardless of where the heads are, the same speed is used to turn the media.

Constant Linear Velocity (CLV): In a CD-ROM drive, the head is on the outside of the disk, the motor runs slower, and when it is on the inside, it runs faster.

Control Bus: It consists of electrical signals (Clock, Read, Write, etc…) that permit the CPU to communicate with I/O devices and the memory.

Control unit: A special module in a digital computer that retrieves and interprets each instruction and ensures that the computer's processing functions are executed with maximum practical efficiency. It extracts program instructions from memory (ROM and/or RAM), decodes and executes them, calling on the *ALU* when necessary.

Control Program for Microcomputers (CP/M): The first Operating System developed for personal computers.

Controller board: A special type of expansion board that contains a controller for a peripheral device (disk drive or graphics monitor).

Co-Processor: It is a special-purpose-processing unit that assists the CPU in performing certain types of operations.

Core storage: A form of high-speed primary storage utilizing magnetic core rings.

Coulomb (Charles Augustin de Coulomb, 1736-1806): The charge (Q) that passes by a point in one second when the current is one ampere. The symbol is *C*.

Cursor (CRT screen): A symbol which indicates where on the screen the next character will be entered. It is a pointer on the display screen.

Cylinder: A collection of tracks which can be referenced by one positioning of the access arm.

»»»»»»»- D -«««««««

Daisy wheel printer: A letter quality, impact printer that consists of a type element that contains raised characters, resembling the structure of a flower, which strikes the paper through an inked ribbon.

Data: Numbers, words and phrases which are suitable for processing in some manner on a computer. It is coded information. Any representations such as characters or analog quantities to which real-world meaning is assigned.

Data banks: Large collections of information stored on auxiliary storage devices.

Data base: A collection of data organized in a manner which allows retrieval and use of that data.

Data base administrator: A person who develops and maintains the data base, controls updating it and often controls who within the business organization has access to the data base.

Data base management systems: Software packages, also called file management systems, that allow use to define files, records within files, and data elements or fields within records in a relatively easy manner, and to provide a convenient method to access, update and create reports from the data.

Data Bus: It transfers the actual data.

Data collection devices: Devices used for obtaining data at the site where transaction or event reported upon takes place.

Data communications: Electronic communication and transfer of data from one computer to another.

Data compression: *Data compression* is storing data in a format that requires less space than usual.

Data integrity: An attribute of data which determines the confidence a user can have in the processing of that data. Data integrity is composed of the three primary elements: data accuracy, reliable data entry and timeliness.

Data item: A piece of data required in the information processing cycle, but normally only useful after being combined with other data items or after a relationship between data items is established.

Data management: Techniques and procedures which ensure that the data required for an application will be available in the correct form and at the proper time for processing.

Data management system: A group of programs, also called a file management system, that performs the actual physical process of reading data from a file stored on disk or tape or writing a record into a file.

Data security and control: Systems and procedures developed for a computer application that allow only authorized personnel to access the data stored in a data base.

Data transfer rate: The time required to transfer data from disk to main computer memory and vice versa.

Daughter card: Any board or card that attaches directly to another board from the expansion slot.

Debugging: The detection, location, and removal of mistakes in a computer program.

Decimal: It is the basic numbering system for counting. The decimal base is 10; there are ten digits in this counting system: 0,1,2,3,4,5,6,7,8 and 9.

Device: It is an electronic unit that is connected to the Bus and either submits commands or carries them out. It is identified by its ID and can act as an initiator or a target.

Dibit (or **Slice)**: It is a group of two bits.

Difference engine: A specific special-purpose digital computer, based on mechanical design principles, developed by Charles Babbage in 1822.

Digital circuits: They are a part of the analog circuits with the exception that the output can be one of two values, called a *bit*: a logical one (1) bit or a logical zero (0) bit.

Digital Transmission: The transmission of data as distinct ON and OFF pulses.

Digital Versatile/Video Disk (DVD): *DVD* uses the same physical form factor as CD-ROM but the logical formats are considerably different.

Digital computer: A computer whose processing functions are based on counting discrete entities rather than measuring continuous variables.

Digitizer pad: A special horizontal sheet divided into a gridlike pattern that allows a person with a stylus to feed pictorial information into a computer and to modify that information.

Diode: It is a natural switch because it operates in either of two logic states: OFF state, its resistance is very high and ON state, it passes current easily.

Diode Transistor Logic (DTL): The *DTL* was developed as a result of some of the problems in the RTL logic such as voltage variation and low fanout.

Direct Memory Access (DMA): The *DMA* is a system resource that assists computer components (video, sound cards, etc.) in communication with the *CPU* and other peripherals within the computer system.

Discrete Cosine Transform (DCT): *DCT* is a technique for representing waveform data as a weighted *sum of cosines*.

Disk cache: It is a mechanism for improving the time it takes to *read from* or *write to* a hard disk.

Disk storage device: A computer storage unit in which binary pulses are stored magnetically on the surfaces of a parallel stack of rotating platters.

Diskette: An oxide-coated plastic disk, also called a floppy disk, which stores data as magnetic spots.

Documentation: Written words, drawings and screen messages which tell a person the exact procedures for all activities associated with a system.

Dot-addressable displays: Display monitors, also called bit-mapped displays, used for graphics.

Dot-matrix printer: An impact printer that uses a movable print head containing small pins that when struck against a ribbon and paper cause small dots that form characters to be printed.

Double Density (DD): Diskettes and drives that record approximately 5,876 bits per inch.

Double Sided (DS): Diskette drives designed so that data can be recorded on both sides of the diskette.

Downloading: Transferring large amounts of data from a host computer to another computer to store on auxiliary storage device.

Downtime: Any interval during which a computer system is inoperative because of a malfunction.

Dual In-line Package (DIP): It is the traditional bug like chip that have **anywhere** from 8 to 40 legs, evenly divided in two rows.

Dual Independent Bus (DIB): An architecture Bus that supports two independent Buses unlike the previous infrastructures (*Socket 1* through *Socket 7*) that had a single Bus architecture; *Socket 8* supports DIB.

Dual Inline Memory Module (DIMM): This memory is similar to *SIMM* but is dual sided, i.e., it can hold twice as many ICs as *SIMM*.

Dynamic Random Access Memory (DRAM): This type of memory needs to use a constant refresh signal to keep the data in memory.

»»»»»»»- **E** -««««««««

E-mail: Electronic mail.

Electrically Alterable ROM (EAROM): It is a memory into which data can be written by a bit or a word, but much more slowly than data readout, also called *Read Mostly Memory*.

Electrically Erasable Programmable ROM (EEPROM): It is similar to the *EPROM* in that by sending a special sequence of electrical signals (pulses) to the *EPROM IC*, the memory circuits can be erased.

Electricity: The electricity is a property of electrons and protons, expressed numerically in coulombs.

Electromechanical relay: A magnetically operated mechanical switch.

Electron: A subatomic particle having a specific negative charge orbiting the nucleus of an atom.

Electronic: Electronic is using or activated by electric current in semiconductors such as transistors, capacitors, and diodes.

Electronics: Electronics is the field of technology that deals with electronic devices such as transistors, capacitors and diodes.

Electronic Delay Storage Automatic Calculator (EDSAC): Developed by Maurice V. Wilkes, it was the first computer using the stored program concept to be operational.

Electronic Discrete Variable Automatic Computer (EDVAC): An early computer developed by John Von Neumann that utilized the stored program concept.

Electronic Numerical Integrator and Computer (ENIAC): The world's first general-purpose electronic digital computer developed in 1946 at the Moore School of Engineering in Philadelphia, PA by J. Presper Eckert and John W. Mauchly.

Emitter Coupled Logic (ECL): The *ECL* is faster than *TTL*. It is used mostly in circuits where high speed (propagation delay time) is essential.

Enhanced IDE (EIDE or ATA-2): The *Enhanced IDE (Expanded IDE or EIDE)* was a newer version of the *IDE* mass storage device interface standard.

Enhanced Small Device Interface (ESDI): The *ESDI* was developed because of the limitation and lack of speed of the *MFM* and *RLL* interfaces.

Erasable Optical (EO): *EO* disks can be read from, written to, and erased like magnetic hard and floppy disks.

Erasable Programmable Read Only Memory (EPROM): It is unlike the *PROM*, *EPROM* is erasable and can be re-programmed, making this type of memory more flexible than *PROM*.

Error: Any discrepancy between a computed, observed, or measured quantity and the true, specified, or theoretically correct value.

Error Checking and Correcting or Error Correcting Code (ECC): *ECC* allows data that is being read or transmitted to be checked for errors and, when necessary, correct it.

Expansion board: Any board that plugs into one of the computer's expansion slots.

Expansion slot: It is an opening in a P C where a circuit board can be inserted to add new capabilities to the computer.

Extended Binary Coded Decimal Interchange Code (EBCDIC): It is a binary code for alphabetic and numeric characters that *IBM* developed for its larger operating systems.

Extended Data Out (EDO) DRAM: It is a type of *DRAM* that is faster than conventional *DRAM*.

Extended Industry Standard Architecture (EISA): In the mid 1980's, the *EISA Bus* was introduced to meet the next generation of faster computer technology.

Extended System Configuration Data (ESCD): The *ESCD* area is a special part of the computer's *BIOS' CMOS* memory, where *BIOS* settings are held.

External storage: Any storage medium outside the computer that can store information in a form acceptable to the computer.

»»»»»»- **F** -«««««««

Farad (Michael Faraday, 1791-1867): The capacitance (C) in which a charge of one coulomb produces a potential difference of one volt. The symbol is **F**.

Fast Page Mode RAM (FPMRAM): The *FPMRAM* is a type of *DRAM* that allows faster access to data in the same row or page.

Feedback: The return of a fraction of the output of a process to its input, to provide for self-adjusting control of the process.

File: A collection of related records treated as a unit.

File maintenance: The periodic process of keeping a file up-to-date by adding, changing or deleting data.

Floating Point Number (FPN): It is a real number such as 1.0, -51.3, 6½ or 4E-2.

Floating Point Unit (FPU): This is a specially designed chip that performs Floating-Point calculations.

Floating-point Operations per Second (FLOPS): It is a common benchmark measurement for rating the speed of computer microprocessors.

Floppy disk: A small, inexpensive magnetic disk unit in which the individual platters are composed of vinyl plastic whose self-lubricating properties permit constant physical contact between the recording heads and the surface of the disk.

Flowchart: A sketch consisting of boxes of various specific shapes interconnected by arrows that provides an overview of the structure of a particular computer program.

Formula Translation (FORTRAN): A symbolic computer programming language used primarily for scientific and engineering applications in which the commands resemble algebraic equations.

»»»»»»»- G - H -«««««««

Gigabyte (GB): A Gigabyte is 10^9 bytes or 1,000,000,000 bytes..

Glitch: A sudden, often unexplained electronic surge that causes problems in an electronic device.

Hard copy: A printed copy of machine output in a readable form.

Hard disk: A hard disk is part of a unit, often called a "*disk drive*", "*Hard Drive*" or "*Hard Disk Drive*" or *HDD* that stores and provides relatively quick access to large amounts of data on an electromagnetically charged surface or set of surfaces.

Hardware: The physical equipment in a data processing center.

Heads: The number of *read/write* heads the media disk uses.

Heuristic: A directed, trial-and-error problem-solving method in which solutions are discovered by consistently evaluating the progress made toward the stated goal.

Hexadecimal: It is another numbering system used often with microprocessors. The hexadecimal (also called hex) is base 16, it uses the decimal numbers 0,1,2,3,4,5,6,7,8,9 and the letters A, B, C, D, E and F.

High-Definition Television (HDTV): It is a new type of television that provides much better resolution than current televisions based on the NTSC standard.

Hobby computer: A small, inexpensive computer purchased for recreational and other home oriented uses.

»»»»»»»- I -«««««««

Icon: A computer-generated image of a familiar object such as computer or trash can. Icon allows inexperienced users to control the computer's activities.

Idle time: The time during which a particular computer is available but is not used.

Information: Data that have meaning to human beings.

Information retrieval: The methods used in recovering specific information from computer storage.

Industry Standard Architecture (ISA): The *ISA* Bus was introduced in the early 1980's with the *IBM PC, XT* and *PC/AT* computer systems.

Input data: The user-supplied data to be processed by a computer.

Input-output device: A computer module used to achieve man-machine communication.

Institute of Electrical and Electronics Engineers (IEEE): An organization composed of engineers and computer scientists.

Instruction: A statement calling for a specific computer operation.

Integrated Circuits (IC): It is a tiny chip of silicon that hosts thousands or millions of transistors and other circuits components (resistors, capacitors, etc...). *Small-Scale Integrated* (*SSI*) circuits (1 to 10 transistors) followed by the *Medium-Scale Integrated* (*MSI*) circuits (10 to 100 transistors), then the *Large-Scale Integrated* (*LSI*) circuits (100 to 5,000 transistors), then the *Very-Large-Scale Integrated* (*VLSI*) circuits (5,000 to 50,000 transistors), the *Super-Large-Scale Integrated* (*SLSI*) circuits (50,000 to 100,000 transistors) and then *Ultra-Large-Scale Integrated* (*ULSI*) circuits (over 100,000 transistors).

Integrated Device Electronics (IDE): IDE is a standard electronic interface between the computer and its mass storage drives.

Interface: The point of contact between two systems.

Internal storage: The addressable storage inside a digital computer directly accessible to the CPU.

International Organization for Standardization (ISO): Founded in 1946, it is a worldwide federation of national standards bodies from some 100 countries, one from each country.

Interrupt: A signal that data has been transferred from main computer memory to a device or vice-versa.

Interrupt priority: The PC processes device interrupts according to their priority level.

Interrupt Request (IRQ): The *IRQ* is a system resource. Every computer has interrupts to get the attention of the *CPU*, as needed.

Inverter: A specific logic gate in which a binary 1 input produces a binary 0 output and vice-versa.

»»»»»»- J - K - L -«««««««

Joule (James Prescott Joule, 1818-1889): The work (W) done by a force of one Newton acting through a distance of one meter. The symbol is *J*. *Joule* is the unit of energy,

Keypunch machine: A keyboard-actuated device that punches holes in a set of cardboard cards.

Kilobytes (KB): A Kilobyte is 10^3 bytes or 1,000 bytes.

Latency: The time it takes for a sector containing data to rotate under the Read/Write Head of a diskette drive.

Least Significant Bit (LSB): It is the right-most bit in a binary number.

Least Significant Digit (LSD): It is the right-most digit of the number because it carries the lowest weight in determining the value of the number.

Light pen: A tiny cylindrical device used in sketching graphical computer instructions on the screen of a CRT.

Link: A transmission channel that connects nodes.

Local Area Network (LAN): A communication network linking various computers or computer terminals so that they can communicate with one another electronically.

Logic gate: It is a circuit with two or more inputs and whose output will be either a high or low voltage.

Logical Block Addressing (LBA): *LBA* is a method used with *SCSI* and *IDE* disk drives. *LBA* is a technique that allows a computer to address a hard disk larger than 528 MB.

Low Insertion Force (LIF) socket: It is an IC socket. The CPU chip is inserted in this type of a socket with a little eccentric force until the chip is fully in place.

»»»»»»»- M -««««««

Machine language: A computer-coding language that can be understood directly by a particular machine without further translation.

Magnetic card: A special plastic card on which data can be stored by selectively polarizing portions of its flat, magnetizable surface.

Magnetic disk: A flat circular plate coated with iron oxide on which data can be stored by selectively polarizing portions of the flat surface.

Magnetic tape: A plastic tape having a magnetic surface for storing data in a code of polarized spots.

Mainframe computer: A large centralized computer, with more processing capabilities than a minicomputer, which is able to store large volumes of data and provide access by numerous users..

Mebibyte: A Mebibyte is 2^{20} bytes or 1,048,576 bytes.

Megabyte (MB): A Megabyte is 10^6 bytes or 1,000,000 bytes.

Megahertz (MHz): 1 MHz is equal to 1 million clock cycles per second.

Memory: The memory in a computer system is a staging area for the CPU.

Memory cache: It is a *RAM* that a computer's microprocessor can access more quickly than it can access regular *RAM*.

Micro Channel Architecture (MCA): *MCA Bus* is an interface between a computer (or multiple computers) and its expansion cards and their associated devices.

Microcomputer: A small, self-contained digital computer with low accuracy and a small price tag.

Minicomputer: A small, self-contained general-purpose desktop computer larger and faster and more versatile than a microcomputer.

Microsecond (μs): A microsecond is 10^{-6} second or One millionth of a second.

Million Instructions Per Second (MIPS): It is a measurement used for the computer's speed and power.

Millisecond (ms): A millisecond is 10^{-3} second or one thousandth of a second.

Mnemonic code: An easy-to-remember computer language code.

Modified Frequency Modulation (MFM): The *MFM* interface controller was developed to resolve the minimal storage and speed of the floppy disk.

Modulation/Demodulation (Modem): A device that converts digital data to analog signals (at the transmitting side), which can be sent over communication channels, then converts back to digital data (at the receiving side).

Most Significant Bit (MSB): It is the left-most bit in a binary number.

Most Significant Digit (MSD): It is the left-most digit because it carries the greatest weight in determining the value of the number.

Motherboard (or mainboard or system board): The principal board that has the most connectors (video, sound, network cards) for attaching devices to the Bus.

Module: A packaged functional hardware unit designed for convenient interconnection with other similar components.

Mouse: A specific type of computer input device. The user slides the mouse along a horizontal surface and depresses one or more buttons located on its top to rearrange the material on the display screen to control various computer operations.

Moving Picture Experts Group (MPEG): *MPEG* refers to the family of digital video compression standards and file formats developed by the ISO group.

MP3: *MP3* is a standard technology and format for compressing a sound sequence into a very small file

Multi-bank DRAM (MDRAM): The *MDRAM* utilizes small banks of *DRAM* (32 KB each) in an array, where each bank has its own *I/O* port that feeds into a common internal *Bus*.

Multimedia: It is the use of computers to present text, graphics, video, animation, and sound in an integrated way.

Multiplexing: The process of transmitting several simultaneous messages over a single communication channel.

»»»»»»»- **N** -«««««««

NAND gate: The NAND gate is equivalent to the AND gate followed by a NOT gate.

Nanosecond (ns): A nanosecond is 10^{-9} second or One billionth of a second.

Network: Any system composed of two or more large computers, PCs or terminals.

Nibble: Can also be referred to as a Binary Coded Decimal (BCD), it is a group of four bits associated together.

Non-volatile storage: Any storage medium that retains its stored contents in the absence of electrical power.

Node: The endpoint of a network, usually PCs with printers attached to a network.

NOR gate: The NOR gate is equivalent to the OR gate followed by a NOT gate.

NOT gate: The NOT gate logic output is equal to the inverse (complement) of the logic input.

»»»»»»»- **O** -«««««««

Octal: It is another number system that is often used with microprocessors. It has a base 8 numbering system, and uses the digits 0,1,2,3,4,5,6,7.

Ohm (George Simon Ohm, 1787-1854): The resistance (Ω) that produces a voltage of one volt when carrying a current of one ampere. The symbol is Ω.

Operating System (OS): An organized collection of software that controls the overall operations of a digital computer.

Optical disk: An optical disk is a storage medium from which data is read and which is written on by lasers.

Optical fiber: A hair-thin glass rod that carries laser-encoded messages from one location to another. If the light beam moves away from the center of the optical fiber, it is reflected or refracted back toward the center.

Optical Character Reader (OCR): A device that reads typewritten, computer-printed, and in some cases hand-printed characters from ordinary documents.

OR gate: The *OR* gate logic output is equal to the *OR* sum of the logic inputs.

Output: The final result of a sequence of data processing operations.

»»»»»»- P -«««««««

PASCAL: A computer programming language which provides statements to encourage the use of structured programming.

Passive backplane: system has a main card with slots mounted on the card and no circuitry on the card.

Password: A value, such as word or a number, which is associated with the user and used to gain access to the computer.

Paper-tape reader: Any device capable of translating the holes in a perforated paper tape into a machine-processable form.

Parity check: An automatic error-detection procedure that uses extra checking bits that are carried along with the numerical bits being processed.

Partition: A partition is a logical organization of the physical disk, (usually hard disk).

PC100 Synchronous DRAM (SDRAM): PC100 SDRAM is a SDRAM that states that it meets the *PC100* specification from *Intel*.

Peripheral device: It is an input/output device that is attached to a computer system internally or externally.

Peripheral equipment: The input-output devices and the auxiliary storage units in a data processing center.

Peripheral Component Interconnect (PCI): The *PCI* Bus is geared specifically to *Fifth-* and *Sixth-Generation Systems*, although the latest generation '486 motherboards use *PCI* as well.

Peripheral Component Interconnect-X (PCI-X): The *PCI-X* is a *PCI* based I/O Bus but improved in performance with increased flow data between CPU and various peripherals.

Picosecond (ps): A picosecond is 10^{-12} second or one trillionth of a second.

Pin-Grid Array (PGA): It is a square chip in which the pins are arranged in concentric squares.

Pixel: Each addressable dot, also called a **picture element**, that can be illuminated on the display screen.

PL/I: A computer programming language designed with some of the computational concepts of FORTRAN and some of the file processing capabilities of COBOL.

Plug and Play (PnP): *PnP* is to create a computer whose hardware and software work together to automatically configure devices and assign resources, to allow for hardware changes and additions without the need for large-scale resource assignment tweaking.

Port: A standard plug and socket with predefined connections, also called a serial interface, needed to transmit a serial stream of bits in and out of the computer.

Primary storage: The main memory unit of a digital computer. The primary storage region usually consists of a high-speed direct-access unit with a moderate storage capacity.

Printed Circuit Board (PCB): It is a thin plastic plate on which ICs, resistors, capacitors and other electronic components are placed.

Printer: A computer output device capable of imprinting alphanumeric characters on papers.

Processing unit: The mechanism used by a computer for altering the input quantities in accordance with the program instructions.

Processor: A computer.

Program: A set of instructions provided to a computer in a machine-readable form.

Programmable Read Only Memory (PROM): It is a type of *ROM*.

Programmed Input/Output (PIO): The *PIO* is a method of transferring data between two devices that uses the computer's main processor as part of the data path.

Programmer: A professional technician who codes instructions in a language amenable to computer solution.

Programming language: Any well-defined language used in preparing instructions for computer execution.

Protocol: A formal set of rules governing the format and relative timing of message exchanges between two communicating devices.

Pseudocode: The logical steps to be taken in the solution of a problem written in the form of English statements.

Pulse: A sharp voltage change.

Punched card: A cardboard card used in data processing operations in which tiny holes at hundreds of individual locations denote numerical values and alphabetic codes.

Punched paper tape: A long strip of paper in which holes are punched to record alphanumeric information for computer processing.

»»»»»»»- R -«««««««

Rambus DRAM (RDRAM): It is a type of memory *DRAM. RDRAM* transfers data at up to 600 MHz.

Random Access Memory (RAM): A semiconductor memory that allows data to be stored and extracted using instructions within a computer program.

Random access device: Any computer storage device in which the time required to retrieve a particular data value is not significantly affected by its physical locations.

Raw data: Data that have not yet been processed or reduced.

Read Only Memory (ROM): A memory in which the information is stored at the time of manufacture. The information is readily available to the user, but it can be modified only with great difficulty.

Read-write head (recoding head): A tiny electromagnet used to record information on a thin, magnetizable film.

Real-time processing: Data processing operations in which the computed results are received so quickly they can be used to influence the operation of the process being simulated.

Recording head (read-write head): A tiny electromagnet used to record information on a thin, magnetizable film.

Refreshed: The process of scanning the CRT screen with an electron beam anywhere from 30 to 60 times per second causing the phosphors to remain lit.

Register: A high-speed device used by the CPU for temporary storage of data values during processing.

Reliability: The quality of freedom from failure. The reliability of a system is usually expressed as the probability that a failure will not occur in a given amount of time or a given amount of usage.

Remote terminal: An array of input-output devices connected to a distant computer by means of telephone lines or other communication channels.

Removable disks: A magnetic disk pack that is removable from the disk drive.

Resistor: It is used to resist (reduce) the flow of electricity (current) in a path by a certain amount. Resistance unit is Ohms (Ω).

Resistor Transistor Logic (RTL): It is built of resistors and transistors.

Response time: The time that elapses between the instant a user enters data and the instant the computer responds to the entry.

Ring network: A series of computers communicating with one another and without a centralized host computer.

Robots: Machines, operating under the control of a computer and related software, that are designed to perform repetitive manufacturing and operational tasks.

Routine: A set of machine instructions for carrying out a specific processing operation.

Run: A single, complete execution of a computer program.

Run Length Limited (RLL): Shortly after the introduction of the *MFM* technology, the *RLL* interface controller was introduced to compete with it.

»»»»»»- S -«««««««

Scanner: Any optical device that can recognize a specific set of visual symbols.

Security: The methods adopted for making valuable property or private information safe from theft or harm.

Semiconductor: A substance, such as selenium, that normally insulates against the flow of electricity but that under certain conditions can be made to conduct the flow.

Sequential retrieval: The retrieval of data from auxiliary storage one record after another based upon the sequence in which the data is stored on auxiliary storage.

Serial interface: A standard plug and socket with predefined connections, also called a port, used to transmit a serial stream of bits in and out of the computer.

Server: A computer, also called a control unit, dedicated to handling the communications needs of the other computers in a Local Area Network.

Settling Time: The time required for the read/write head to be placed in contact with the disk.

Single Instruction Multiple Data (SIMD): A process that allows an instruction to perform the same function on many pieces of data which cuts down the repetitive loops of instructions that consume vast amounts of time to execute.

Single In-line Memory Module (SIMM): It is a *PCB* that consists of nine or more packaged as a single unit.

Single In-line Package (SIP): It is an IC that has just one row of legs in a straight line like a comb.

Single Inline Pinned Package (SIPP): The *SIPP* is very similar to a *SIMM* except it has pins on the bottom of the module that interface to the motherboard instead of an edge connector like the *SIMMs*.

Single density: Diskettes and drives that record approximately 2,768 bits per inch.

Single sided drives: Diskette drives designed so that data can be recorded on only one side of the diskette.

Slice (or dibit): It is a group of two bits.

Slide rule: A hand-operated computing device consisting of a ruler with a sliding center section, both of which exhibit logarithmic scales.

Small Computer System Interface (SCSI): The *SCSI* is not a disk interface, but a systems-level interface. *SCSI* is not a type of controller, but a Bus that supports multiple devices.

Slot 1: Intel's proprietary, it is the *form factor* for *Intel's Pentium II* and *Pentium III* CPU.

Socket 3: It is the *form factor* for fifth-generation CPU chips such as '486 CPU generation from *Intel*, *AMD* and *Cyrix*.

Socket 7: It is the *form factor* for fifth-generation CPU chips such as *Intel's Pentium* series, *AMD K6* series and *Cyrix* CPU.

Socket 8: It is the *form factor* for *Intel's Pentium Pro* CPU.

Soft-sectored diskettes: A diskette on which the number and size of sectors is determined by the formatting of the diskette.

Software: All the instructions and documents needed for guiding the operation of a computer.

Software encryption: The encoding or decoding of computerized data using programming techniques rather than hardware devices such as scramblers.

Solid state device: A non-vacuum electronic device fashioned from semi-conducting materials that performs some of the functions of a vacuum tube.

Storage: A computer memory.

Storage capacity: The maximum quantity of information that can be retained in a particular storage device.

Storage device: A device that can accept information, hold it, and deliver it on demand at a later time.

Static Random Access Memory (SRAM): It is static because the information stored in this type of memory does not need a constant update or refresh.

Superscalar: A CPU that can execute more than one instruction per clock cycle.

Synchronous Computer: A digital computer that requires the same amount of time to complete each processing operation regardless of its intrinsic complexity.

Synchronous Graphic RAM (SGRAM): The *SGRAM* is a type of *DRAM* used increasingly on video adapters and graphics accelerators.

»»»»»»»- **T** -«««««««

Tape density: The number of characters or bytes which can be stored on an inch of tape.

Tape drive: A device that uses a magnetic tape to read or write data information.

Timeliness: The data to be processed has not lost its usefulness or accuracy because time has passed.

Transistor: It is used in a circuit as a switch. A small voltage is applied at one pole that switches a larger voltage at the other poles (ON or OFF).

Truth table: A systematic tabulation associated with a binary circuit listing all possible combinations of input values and indicating, for each combination, the resulting outputs.

Trackball: A plastic sphere mounted in a socket so that it is free to rotate in any direction.

Touch screen: A simple computer input device whereby the user controls the functioning of a digital computer by pressing a finger to the screen at specific locations.

Throughput: A measure of a system's efficiency; the rate at which useful work can be accomplished.

Token: A string of bits that constantly travels around the network.

Track: A very narrow recording band forming a full circle around the diskette.

Tracks per inch: A measurement of the number of tracks a diskette drive can record.

Transistor Transistor Logic (TTL): The *TTL* displaced the *DTL* because of the speed and low impedance in both logic 1 and logic 0 output states.

Turnaround time: The interval between the submittal of a computer job and the time it is returned in finished form to the user.

Twisted pair wire: Common telephone wire cord.

»»»»»»- **U** -«««««««

Ultra-ATA: Also called *Ultra-DMA, ATA-33,* and *DMA-33,* supports multiword *DMA* mode 3 running at 33 Mbps.

Unconditional transfer: A transfer of control in a computer program that is not dependent on the outcome of a previous computation.

Universal Automatic Computer (UNIVAC I): The first electronic computer dedicated to business applications.

Universal Product Code (UPC): A specific binary code used in grocery stores and other retail establishments whereby the ten decimal digits are represented by patterns of black and white strips that can be decoded by a computer equipped with optical scanners.

Unix: An Operating System (OS) developed by Bell Labs for use on minicomputers but which has been modified to run PCs.

Uploading: The transmission of data from files on the PC to data bases on the host computer.

User: Anyone who utilizes the services of a data processing center.

User friendly: Describing a computer program specifically designed to provide prompts, pictorial information, and other aids to help untrained users operate it successfully.

Unicode: It is an entirely new idea in setting up binary codes for text or script characters. It is a system for the interchange, processing, and display of the written texts of the diverse languages of the modern world.

»»»»»»- **V** -«««««««

Video Electronic Standard Association (VESA): The *VESA* or *VESA local Bus* or *VESA VL Bus* or *VLB,* was introduced in 1992 to allow high-speed connections between peripherals.

Video RAM (VRAM): The *VRAM* is a special-purpose memory used by video adapters.

Voice input: The ability to enter data or issue commands to the computer with spoken words.

Voice output: A form of output consisting of spoken words which are conveyed to the computer user from the computer.

Voice recognition: A system which understands human speech regardless of the speaker or the words which are spoken.

Voice synthesizer: A type of voice generation which can transform words stored in main computer memory into speech.

Volatile: A storage system in which the information vanishes if the power is turned OFF or temporary interrupted.

Volt (Alessandro Volta, 1745-1827): The potential difference between two points on a wire carrying one ampere of current when the power dissipated between the points is one watt. The symbol is *V*.

»»»»»»»- W -««««««

Watt (James Watt, 1736-1819): The power (P) required to do work at a rate of one joule per second. The symbol is *W*.

Winchester disk: A type of hard, fixed disk often used with PC.

Windows RAM (WRAM): The *WRAM* enables a video adapter to fetch the contents of memory for display at the same time that new bytes are being pumped into memory.

Word: A group of characters that have one addressable location and are treated as a single unit by the computer.

Write-Once, Read-Many (WORM): *WORM* disks can be written on once, after that, the *WORM* disk behaves just like a *CD-ROM*.

»»»»»»»- X - Y - Z -««««««

Zero Insertion Force (ZIF) socket: It is an IC socket that allows the insertion and removal of an IC chip without special tools.

BIBLIOGRAPHY

The followings books will provide further reading on computer systems and architectures. These books contain detailed references for the subject in depth and more specific.

- Tocci, Ronald J., *Fundamentals of Pulse and Digital Circuits, Second Edition.* Columbus, Ohio: Charles E. Merrill Publishing Company. 1977.
- Tucker, Allen B., Bernat Andrew P., Bradley W. James, Cupper Robert D., Scragg Greg W., *Fundamentals of Computing I, Logic, Problem Solving, Programs and Computers, Pascal Edition Revised.* McGraw-Hill, Inc. 1994.
- Mandell, Steven L., *Computers and Information Processing, Fifth Edition.* West Publishing Company. 1989.
- *AMD*, the *AMD* Logo, and the combinations thereof are trademarks of Advanced Micro Devices, Inc.
- *Microsoft*, *DOS* and Windows are registered trademarks of *Microsoft Corporation*.
- *Pentium* is a registered trademark, and *MMX* is a trademark of *Intel Corporation*.
- Other product names used in this publication are for identification and reference purposes only and may be trademarks of their respective companies.